GEOGRAPHY OF INDIA

GEOGRAPHY OF INDIA

PRITHVISH NAG
SMITA SENGUPTA

CONCEPT PUBLISHING COMPANY, NEW DELHI 110059

ISBN 81-7022-384-9 (H.B)

First Published 1992

Second Reprint 1999, 2002

Published and Printed by
Ashok Kumar Mittal
Concept Publishing Company
A/15-16, Commercial Block, Mohan Garden,
NEW DELHI - 110059
Phone: 5648039, 5649024
Fax : 091-(11)-5648053

Dedicated
to
Professor S.P. Chatterjee

PREFACE

This book is the outcome of my close association with Prof. S.P. Chatterjee, the doyan of Indian geography and cartography. I was closely associated with him regarding his book on *Economic Geography of Asia*. The third section of this book was on India. Nevertheless, Prof. Chatterjee inspired me to work on a fresh book on the *Geography of India*. Under his guidance, I started the work. We both were co-authors of the proposed volume. Four chapters were written for this purpose, but the progress was not satisfactory. First, it was due to my absence from NATMO, Calcutta for about a year and half, and later Prof. Chatterjee's health did not permit him to take the strain of going through the draft chapters. However, the idea of writing a book on the geography of India was always in mind. The earlier scheme was redrafted, made shorter, and then a fresh attempt was made to bring out this book. Obviously, under these circumstances, the readers will find the reflections of Prof. Chatterjee's ideas in this book which is dedicated to him.

When I was a student in the Banaras Hindu University, I had seen Prof. R.L. Singh working untiringly for his book on *India : A Regional Geography*. Both these geographers have greatly inspired me to work on the present book. Further, perhaps there cannot be a better place than the National Atlas and Thematic Mapping Organisation, Calcutta to work on such a book. Its rich library, collection of maps and data, and the most important, the nature of work provides one of the best opportunities in the country for this purpose.

Fortunately, I met Dr. Smita Sengupta who after completing her Ph.D. from Gujarat University, Ahmedabad came back to Calcutta and wanted to take up a challenging work in geography. She accepted my invitation of becoming the co-author of this volume and took great pains in completing it.

The aim of writing this book was to cater persons who are more serious in knowing about the geography of India. It is not an exhaustive book on the topic. Books have been written on each theme included here. The objective is to introduce these themes in a geographical perspective. The selected bibliography i.e. references has been included in each chapter as well as at the end of the book so that any body who is interested in knowing more details will get some guidelines. The language of the book has been made a simple as possible with a view to making it useful for everybody. People may find that the contents meet the requirements of the graduate courses in geography of India as well.

Separate chapters on important theme has been included. Some unconventional chapters have also been included in order to provide a glimpse of the new issues emerging over the national scene. These chapters are on environment and Indian Ocean. Since no book on the geography of India has been produced recently, special attention has been made to use the recent data. Attempts have been made to include in each chapter at least a representative table along with a map. The cartography for this volume was done by Shri Sunil Kumar Dey.

This work would not have been completed without the active support of the members of the NATMO library, Dr. G.K. Dutt and Shri A.K. Kundu, both former Directors, Dr. B.K. Roy, Director and Shri A.K. Sengupta, former Joint Director, Shri R.K. Banerjee, Joint Director and members of my section in NATMO. Prof. P.K. Sircar and my friends and colleagues all over the country inspired and helped me in this work. I have been particularly encouraged by Shri Y.S. Rajan, Advisor, and Shri S. Ghose, Joint Secretary, both in the Ministry of Science and Technology to take up challenging work like this. We are grateful to them. We are particularly indebted to the members of our families for supporting us in our initiatives. Finally, its my pleasure to get the fourth book published from the Concept Publishing Company. Shri Naurang Rai and Shri Ashok Kumar of this company stood by us in our endeavour.

PRITHVISH NAG

CONTENTS

LIST OF TABLES

LIST OF MAPS AND FIGURES

1

INTRODUCTION

India is a concept. India is an experiment through ages. India is an exception. India is India — no parallel example exists. Spiritual India cannot be disassociated with the geographical India. Its cultural eminence can be traced by its ancient literature, scientific achievements in the past and its religio-spiritual contributions to the world. Its passage through ages and under different rulers could not change this character. It will be apparent from the historical and cultural background of this country that the spiritual India has contributed a lot than the India as a territorial unit which we understand today. Obviously, the geographical concept of India is only a part of the philosophical concept of India. Though this country is a distinct geographical entity with well marked mountains and sea (Appendix I).

The Indian Union extends from 8°4′ to 37°6′ north and 68°7′ to 97°25′ east (the southern extent actually goes up to 6°45′ to cover the last island of the Nicobar group of islands) covering an area of 3,287,782 km². Its land frontier is about 15,200 km and coastline is 6,100 km. It extends from the snow covered Himalayan mountains in the north to tropical rain forests in the south. The country is surrounded by Pakistan, Afghanistan, China, Nepal, Bhutan, Myanmar and Bangladesh. Bay of Bengal, Indian Ocean and Arabian Sea are touching the Indian coast. Maldives or Male, Sri Lanka and Indonesia are very close to the India's mainland or its islands (Map 1.1). In fact the Indian Ocean has never been a barrier between the countries and cultures. History clearly reveals that this ocean was responsible to bring people closer. Transcontinental trade, exchange of population among the littoral states etc. were common in the past. Obviously, the geography of this region has been influencing the historical events which in turn have been responsible to give the shape of modern India.

The geographical position of India made it possible to keep contacts with other countries of the world. It was from here where the

MAP 1.1: India: Administrative Set-up

Hindu and Buddhist cultures and religions spread over Asia and the Far East. There are a number of temples with *slokas* from *Ramayana* engraved on the walls in Thailand. Bali in Indonesia was the seat of Hindu kingdom. There was a close contact between the regions and countries in the Far East due to the spread of Buddhism. It is also claimed that there was a close contact between west Indian coast and east African coast. Further in the west, there were established trade routes with Egypt, Babylon, Persia and Rome. All these contacts helped India to propogate its culture to the other regions and also to integrate some of their cultures with the main stream of the Indian culture.

There has been waves of settlers from the northwest who were lured by the riches of India in general and the Great Plains in particulars. These settlers were able to form nations and principalities and slowly assimilated in the national main stream. India as it stands today is a mixture of races, cultures, peoples and ethnic groups. The diversity in the unity is not only due to the physical variations within a geographical unit, but also due to conglomeration of cultural variations in a political unit. The diversity of all kinds do not exactly fit in the western definition of 'nation-state'. However, India with democratic, federal and socialistic form of government, is the biggest experiment now being done on earth in political science. India marches forward in every field of technological advancements everyday along with the poverty, over population and lacking social amenities.

Its population of 685.2 million (1981) is distributed in 32 states and union territories, or in about 4 thousand towns and cities and about 6 lakh villages. The overall density is 221 persons per square km. The annual growth rate of population has 2.26 per cent which is increasing despite of the best efforts by the government in promoting family planning. About 76.7 per cent of the population live in rural areas and are mostly dependent on agriculture for their livelihood. However, the percentage share of agriculture in the net national product has gone down from 58.9 in 1950-51 to 35.4 in 1985-86. The net area sown is about 46 per cent which has remained more-so-ever constant since 1970-71. Similarly, the total cropped area is around 57 per cent since 1975-76. During this period the agricultural science has undergone drastic changes. The green revolution followed by white revolution has made their impacts. Foodgrains are now exported. Recently, the surplus production of oil-seeds has stopped import of edible oil. These changes have been possible only due to efficient management of land, fertilisers, machines and funds at all levels. However, much development is desired in order to cope with the teeming population. The small size of operational landholdings will have to fit in the overall transformation strategy.

The technological scenario is also changing at a faster rate. This factor has played an important role not only in development but also contributed to the unity of the country which is in the form of electricity generation and supply, railways, communication (radio and TV) etc. About 19 per cent of total consumption of electricity is in agricultural sector. The production of tractors has gone over 80 thousand per year, and import has been stopped since 1977-78. Power tillers over 4.2 thousand in a year are now produced here. Installation of electric and diesel pump-sets are in progress. About 64 per cent of the villages are electrified.

The progress of the country is phased according to the Five Year National Development Plans, starting from 1951. The 7th Plan has just ended in 1990. The plan outlay has increased from Rs. 1960 to Rs. 180,000 crores. The emphasis has always been on the agriculture, industry and energy sectors. However, diversification in governments spending and activities has taken place in a big way. Union departments and ministries on science and technology, space, electronics, environment, non-conventional energy, ocean development have been established. In addition, there are several national laboratories and institutions under the Council for Scientific and Industrial Research, Indian Council for Agricultural Research, Indian Council for Medical Research, Oil and Natural Gas Commission, Indian Institute of Technology and University Grants Commission. India has already shown its capabilities in space research and applications, rocketary, Antarctica expedition, computer software development, and atomic energy. It has developed substantial industrial base, and the food stock is sufficient enough to face any eventuality. For example, the drought of 1987 was less severe than the earlier droughts. The rate of inflation has gone down to moderate 5 per cent.

India has certainly gained stability and maturity in the past four decades of post-Independence era. It has absorbed shocks of all sorts, and emerged as a leader in its own right. It has all the potentialities, and the scenario of twenty first century will remain unpredictable. Options and opportunities can take this country to take up any challenge or role in global affairs.

REFERENCES

Chatterjee, S.P. (1965), Physiography, ***Gazetteer of India***, Government of India Publication, New Delhi (Chapter I).

Chatterjee, S.P. (1984), ***Economic Geography of Asia***, Allied Book Agency, Calcutta.

Mukerji, A.B. & Aijazuddin Ahmad (eds.) (1985), ***India: Culture, Society and Economy***, Inter-India Publication, New Delhi.

NATMO (1982), ***National Atlas of India***, National Atlas and Thematic Mapping Organisation, Calcutta.

NATMO (1985), ***Glimpses of India***, National Atlas and Thematic Mapping Organisation, Occasional Monograph 2, Calcutta.

Publication Division (1989), ***India 1988-89— A Reference Manual***, Ministry of Information and Broadcasting, Government of India, New Delhi.

Singh, R.L. et al. (eds.) (1971), ***India: A Regional Geography***, National Geographical Society of India Publication, Varanasi.

Singh, R.L. (1979), Evolution of clan territorial units through land occupance in the Middle Ganga Valley, ***N.G.J.I.***, Vol. 20(1), pp. 1-19.

Spate, O.H.K. & A.T.A. Learmonth (1967), ***India and Pakistan***, Metheun.

2

HISTORICAL BACKGROUND

It is imperative to study the historical background of India in order to understand an ancient country like this. The role of geography in influencing the historical processes cannot be neglected. This is all the more important for understanding the genesis and assimilation of people, cultures and races over time and space. The classical Sanskrit literature, the Vedas, Upanishads, epics such as *Ramayana* and *Mahabharata, Puranas,* contributions of Panini, Patanjali and Kautilya, dramas and poems of Kalidasa and the like contain description of the geographical facts which helps in explaining the moulding of the Indian culture. These evidences present a systematic record of the cultural unity of India (Map 2.1).

Pre-historic Period

There are archaeological evidences suggesting traces of human activities and culture during the pre-historical period. The earliest evidences have pointed the possibility of the rich Harappan Culture (*c.* 2300 B.C.) which is evident from the Indus valley excavations or other sites of the same period. These sites are mostly in the western part now either in Pakistan or in Rajasthan and Gujarat of India. This culture was based on cities, such as Mohen-jo-daro, Harrappa, Kot Diji (in Sind), Kalibangan (in Rajasthan), Rupar (in Punjab) and Lothal (near Ahmedabad in Gujarat). The sources to sustain this culture were surplus produce of the country side, and trade. The layout and other evidences indicate about the civic planning and the organisation of these cities.

Vedic Period

In the Vedic period, about beginning of the third millennium B.C., this region known as Jambudvipa, was a geographical entity and was

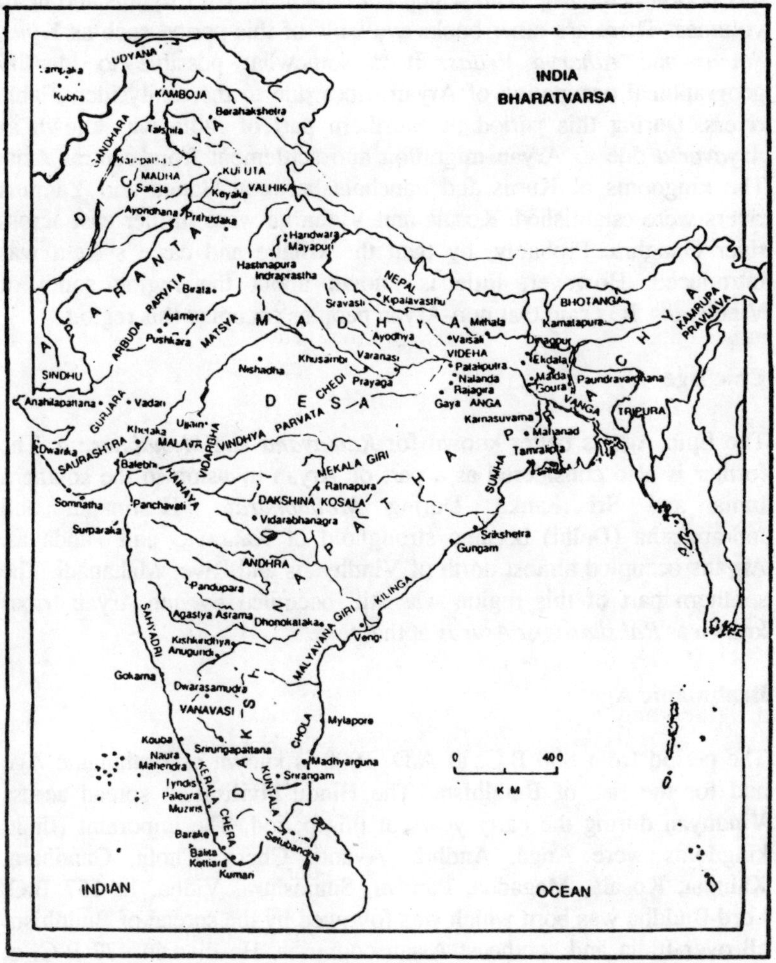

Source: S.P. Chatterjee (1976 to 1982), National Atlas of India

MAP 2.1: India: Bharatvarsa

divided into five regions which is somewhat similar to the modern concept of the regional division of the country. The earliest evidence of this period is the *Rig Veda* which is available in 1017 hymns and in 10 volumes. There are other books available of this period such as *Yajus*, *Samas* and *Atharva Vedas*. It is somewhat possible to identify geographical occupation of Aryan tribes due to the easily identifiable rivers. During this period the northern part of India was known as *Aryavarta* due to Aryan migration and settlement from central Asia. The kingdoms of Kurus and Panchals between Ganga and Yamuna rivers were established. Kosala and Vidharva were further east across river Ghaghra. Probably, by then the village and caste system was introduced. However, little is known about the region south of Vindhyans. It is said that non-Aryan peoples occupied this region.

Epic Age

The Epic Age is better known for *Ramayana* and *Mahabharata*. The former is also considered as a part of Aryan invasion to the southern India and Sri Lanka. During *Mahabharata*, Hastinapur and Indraprastha (Delhi) became stronghold of Kauravas and Pandavas. Aryans occupied almost north of Vindhyans and River Mahanadi. The southern part of this region was still occupied by non-Aryan tribes known as *Rakshasas* or *Asuras* at that time.

Brahmanic Age

The period from 800 B.C. to A.D. 1000 is known as Brahmanic Age and for the rise of Buddhism. The Hindu civilisation spread across Vindhyan during the early years of this period. The important Hindu kingdoms were Anga, Andhra, Avanti, Chera, Chola, Gandhara, Kalinga, Kosala, Magadha, Pandya, Saurashtra, Vidha. In 557 B.C. Lord Buddha was born which was followed by the spread of Buddhism all over India and southeast Asian countries. He died in 477 B.C. at Kusinagar. On the other hand the attacks from the west continued in different periods. Darius invaded Punjab in 500 B.C. and Alexander the Great defeated Porus in the battle of Jhelum in 326 B.C. New dynasties were also established, such as Nanda dynasty (370 B.C.) and Maurya dynasty (321 B.C.), both in Magadha.

This period was also known for its literary contributions and for sowing the seeds of Buddhism. *Sutras*, *Vedangas*, laws of Manu,

Nyaya by Gautam, *Purva Mimansa* by Jaimini, *Sankhya* by Kapila, *Uttar Mimansa* by Vyasa, *Vaiseshika* by Kanada, *Yoga* by Patanjali were composed. Megasthenes visited Chandragupta's court in 305 B.C. Also, Buddhist councils were founded in Rajagriha and Vesali, both now in Bihar. *Arthashastra* written by Kautilya, an advisor in Chandragupta's court, describes the nature of administration, taxation land, economics during this period. For the first time, the Aryavarta was established from Hindu Kush to Himalaya, and beyond Vindhyan touching seas in the east and the west.

Buddhist Age

The actual commencement of Buddhist Age may be considered from the coronation of Ashoka as the king of Magadha (263 B.C.). In 256 B.C. he conquered Kalinga and spread Buddhism which also became the state religion. Buddhist missionaries were sent abroad, including Sri Lanka. With the death of Ashoka in 232 B.C., the Magadha kingdom declined. Later, in 27 B.C. Andharas took over Magadha. Mauryan kingdom under Ashoka spread in all directions, from Bengal to the west of the Indus. Only the southern part of India was under Chola kings.

During the early years of *Anno Domini* (A.D.), Kadphises, king of the Kushans occupied the north-western part of the country (A.D. 50-85). His successor Kadphises II expanded this kingdom in northern India. In A.D. 125 Kanishka succeeded him and expanded the Kushan empire from the borders of Turkestan to the Vindhyans. With the death of Kanishka, the Kushan kingdom declined. In the south of Vindhyas, in quick succession, a lot of changes took place during A.D. 100-250. By the end of this phase the Andhra and Kushan empires fell, Magadha power revived and Pallava rose to prominence.

Gupta Empire and Puranic Age

The period A.D. 300-450 is considered to be the golden age of Hinduism. The art and learning, Sanskrit literature, *Puranas*,were some of the indicators of Hinduism. Chandragupta came to power and in A.D. 320 he became *Chakravarti* (paramount) emperor which also marked the beginning of Gupta era in the Indian history. He was succeeded by Samudragupta (A.D. 326) who expanded the kingdom further in the south. He was succeeded by yet powerful Chandragupta

Vikramaditya (A.D. 375-413). The famous poet Kalidas was in his court. The Chinese traveller, Fa Hian, visited during this period. Vikramaditya also expanded the kingdom in western India. This was followed by the Kumargupta's reign (A.D. 413-455) which was attacked by Chalukyas from the south and Huns from the northwest. Skandagupta took over in A.D. 455 and with his death in A.D. 480 the Gupta empire fell.

In the beginning of the seventh century, Harshavardhana in Thaneswar (north of Delhi), became supreme power in northern India, and Chalukyas in southern India. The former kingdom was overthrown by Chinese (A.D. 650), while the latter kingdom expanded by capturing Pallava kingdom and subduing Cholas and Panchyas. This period is also famous for *Paigambar* Mohammad (A.D. 570-632) and for the visit of Hsuang-Tsung to India.

There are evidences indicating wellorganised flourishing kingdoms in southern India. The description by Megasthenes and Kautilya, and by the author of the *Periplus of the Erythraean Sea* highlight the polity, culture, trade, economy of the southern kingdoms including Tamil and Singhalese kings.

Rajput Age

Attacks from the northwest and in fighting among the local kings continued. Kingdoms changed their rulers by both these processes. Bengal, which had been difficult to rule, came under Pal dynasty (A.D. 800-1060). Mahmood of Ghazni was overrunning the plains. He took over Lahore in A.D. 1001, Peshawar in A.D. 1009, Thaneswar in A.D. 1014, Kanauj in A.D. 1018 and attacked the famous Somnath temple in A.D. 1025. This was the beginning of the Muslim rule in India.

Another attack from the northwest by the Md. Ghori. After destroying Ghazni in A.D. 1149, he came to Punjab in A.D. 1186, but Prithviraj, king of Ajmer, defeated him near Thaneswar in A.D. 1191. Two years later, Prithviraj was defeated and killed by Ghori, and the northern India came under his rule. By then Sen dynasty came to power in Bengal (A.D. 1050-1194).

Afghan Rule

In 1206 Mohamed died and Kutubuddin became the Emperor of India and founded the Slave dynasty. Their territory expanded in the northern India, and ended in A.D. 1290 with the attack by Jalaluddin Khilji who sowed the seeds of Khilji dynasty (A.D. 1290-1320). During his period, the Mughals started attacking India but were unsuccessful. Khilji rule extended to Rajputana, Maharashtra, Telangana while Cholas ruled south of Vindhyans. From time to time, Chalukya, Hoyasala, Yadava all in the South became powerful and got involved in in-fighting.

Khilji dynasty was succeeded by Tughlak dynasty in Delhi. Bengal had independent Muslim kingdom (A.D. 1340), and Vijayanagar in the south had a Hindu kingdom (A.D. 1343). After the death of Firoz Tughlak (A.D. 1351-1388), Gujarat, Sind, Malwa and Khandesh became independent of Delhi's Tughlak rule. During this period Bahamani kingdom in southcentral India became powerful and established rule over Vijayanagar and Telangana (A.D. 1397-1422).

In A.D. 1398 Timur, again from northwest, invaded India and sacked the rule in Delhi again. Following this, the Syed dynasty rule was short-lived in Delhi (A.D. 1414-1450) which was succeeded by yet another short lived Lodi dynasty (A.D. 1451-1517). Kingdoms south of Vindhyans such as Bijapur, Golkunda and Ahmadnagar became independent. The Sikh Guru, Nanak, was born in A.D. 1469. The another most important event of this era was the discovery of the sea route to India by Vasco-da-Gama (A.D. 1498). Portuguese reached Calicut in A.D. 1500 and Cochin in A.D. 1503. New trade and commerce activities started with the arrival of Europeans which lead to the colonisation of India and other countries and as the history of the later periods will reveal, such an event had a long lasting impact in this part of the world.

The year A.D. 1526 is famous for two reasons. The Bahamani empire inbetween Vindhyan and Deccan fell and Babar, Sultan of Kabul defeated Ibrahim Lodi in the first battle of Panipat, near Delhi thereby became the first Mughal emperor in Delhi. But he was short lived. After his death in 1530 his son Humayun became the emperor. In the same year, the Hindu ruler Krishnadeva died and the Vijayanagar empire in the south declined. Following this, several events took place in quick succession. Sher Shah captured Bihar and Bengal (A.D. 1538) and defeated Humayun at Buxar in Bihar in A.D. 1539. Further in following year Humayun was again defeated at Kanauj and fled. Same year (A.D. 1540) Sher Shah became emperor of Delhi but was killed after 5 years. Humayun re-established Mughal empire in A.D. 1555 and died in the following year.

Mughal Rule

The actual Mughal rule is considered from the Akbar the Great, son of Humayun. In the second battle of Panipat he defeated Hemu in A.D. 1556. A lot of infighting went on between the kings, north and south of Vindhyans. Ultimately, the empire extended from Orissa, north of Vindhyans up to Kabul, while Ahmadnagar to Berar in the south of Vindhyans was divided in 15 subas. Kashmir (A.D. 1587), Sind (A.D. 1592), Kandahar (A.D. 1594), Ahmadnagar and Khandesh (both A.D. 1600) became parts of the Mughal empire. Raja Todar Mall efficiently organised the revenue collection. There was all round development during the Akbar's rule (A.D. 1556-1605).

During this period, the activities in Europe were initiated which affected India in the years to come. The English East India Company (A.D. 1600), Dutch East India Company (A.D. 1602), and French East India Company (A.D. 1604) were established with the patronage of their respective governments. There were battles between these powers, including that of Portuguese to get either a foothold on the Indian soil or to carryout profitable business.

After the death of Akbar his son Jahangir became the next Mughal emperor. Captain Hawkins visited Agra in A.D. 1609 and four years later they were allowed to establish factories. In 1627 Jahangir died and in the same year Shivaji, the Maratha ruler was born. Shah Jahan was the next Mughal ruler (A.D. 1627-1658) who later made Taj Mahal in memory of his queen Mumtaj Mahal. The European powers were getting stronger and Maratha kingdoms were expanding their territories. In A.D. 1657 Shah Jahan fell ill and battle among the four sons for the throne started. Finally, Aurungzeb succeeded to the Mughal throne (A.D. 1658-1707). He expanded the kingdom up to river Krishna but Marathas were constant source of botheration. Shivaji again became an independent ruler in A.D. 1674. Aurungzeb persecuted Hindus and Sikhs. Inbetween A.D. 1676 to 1708. Guru Gobind Singh formed Sikh confederation in the northwestern India.

During the period A.D. 1683-1689 Aurungzeb continued to attack Marathas. Bijapur and Golkunda were annexed in 1686 and 1687 respectively. But Marathas conquered Deccan later. The principalities south of Vindhyans, particularly Marathas, were always a problem for the Delhi rulers. Aurungzeb had to come to south in his last days and finally died at Ahmadnagar in A.D. 1707. During his reign, East India Company started to establish its dominion on the Indian soil. The later

Moghul kings were weak and in A.D. 1717. English were given trading rights in Bengal and Calcutta.

Maratha Ascendancy and British Rule

As the ruler in Delhi became weaker, the territories in the east and south became independent and this opportunity was exploited by the European powers. Marathas extended their sovereign over Deccan replacing Mughals. Alongwith Nizam, they overthrew the Mughal king Farukh Siyar. Both these kingdoms grew steadily while with the attack of Nadir Shah of Persia the Mughal emperor totally collapsed (A.D. 1738). Indore, Gwalior and Baroda became Independent states. Change of sovereignty over parts by the European powers continued. During the mid-18th century French became powerful. They took over Madras (A.D. 1746), Pondicherry (A.D. 1748), and in 1750 Deccan was under their rule. At the same time, Clive of the English East India Company came to power and took Trichinopoly, south of River Cauveri (A.D. 1752). The period following this had been very eventful. Every year up to the end of 18th century, the events had some bearing on the course of the Indian history. A different result from any one of these events would have changed the course of the country. Some of these important event were: (a) Second Carnatic war in the south, (b) capture of Arcot in the west of Madras by Clive, (c) surrender of French at Srirangam, (d) capture of Calcutta from the English by Siraj-ud-Daula, Nawab of Bengal, followed by the battle of Plassey, (e) French driven from Northern Circars (Orissa and Andhra Coast) and Pondicherry, (f) third battle of Panipat, near Delhi, (g) defeat of Nawab of Oudh at Buxar, (h) Marathas defeated Haider Ali of Mysore, (i) birth of Maharaja Ranjit Singh in Lahore, and (j) administration of East India Company under the Crown.

By the end of 18th century, the English took over Tanjore, Surat, Carnatac and parts of Oudh. Madras became a province. War and battles between the local rulers and the English continued. At the time of Wellesley, part of present Uttar Pradesh and Bihar, Bengal plains, Northern Circars, Carnatac up to west Indian coast were under British rule. Treaties with several local rulers were made. In A.D. 1819 Ranjit Singh took over Kashmir. With British rule getting firm in India, several reforms were initiated, such as legal reforms. Hindu College at Calcutta was opened in A.D. 1817. After Queen Victoria ascended to the British throne (A.D. 1837), a lot of changes took place in western and north western India. Afghan became powerful, defeated British army in Khyber pass (A.D. 1842), later subdued. Sind was annexed in the following year. Sikhs were defeated and forced to have treaty with the British.

First War of Independence

The British annexation policy continued. In a series territories were dissolved: Satara in western India (A.D. 1848), Punjab (1844), Bassein, Prome, Pegu and Rangoon (A.D. 1852), Nagpur, Jhansi and Berar (A.D. 1853), and Oudh (A.D. 1856). At the time of the first war of independence (A.D. 1857), the plains of Indus and Ganga, Eastern coast, Bombay, Satara, and Nagpur were British territories. In the year the Mughal Empire in Delhi ultimately fell. Following this war, the crown assumed direct administrative control over India (A.D. 1858). Twenty years later Queen Victoria became Empress of India. On one hand, universities were opened; while on the other, famine, plague, earthquake and wars created human problems of all sorts. The annexation policy continued and territories were redefined. Assam became a separate province (A.D. 1874). North West Frontier Province was formed. Bengal was partitioned in 1904 and again reunited in 1911. Same year capital was transferred to Delhi from Calcutta, and the new provinces of Bihar and Orissa were carved out.

Indian National Congress was formed in A.D. 1885, and after the return of Mohandas Karamchand Gandhi, a British educated lawyer, from South Africa who took the lead which resulted into the Independence of the country in A.D. 1947. Mahatma Gandhi was later known as the father of the nation. From A.D. 1900 to 1935, a lot of changes took place which are related to viceroys, World War I, and administrative changes through several acts. By 1937, the provincial autonomy act came into force and provision for all India federation was made. Same year Burma was separated from India. Just before the beginning of the World War II (A.D. 1939-45), India under the Crown consisted territories of British and Indian states (Map 2.2). The former territories were having Bengal, Madras and Bombay presidencies, apart from other types of administrative set up. During this time, the communal scene became worst and all the efforts to bring harmony fizzled out. India and Pakistan became independent in 1947. In the following year Burma and Sri Lanka also became free of the British Rule.

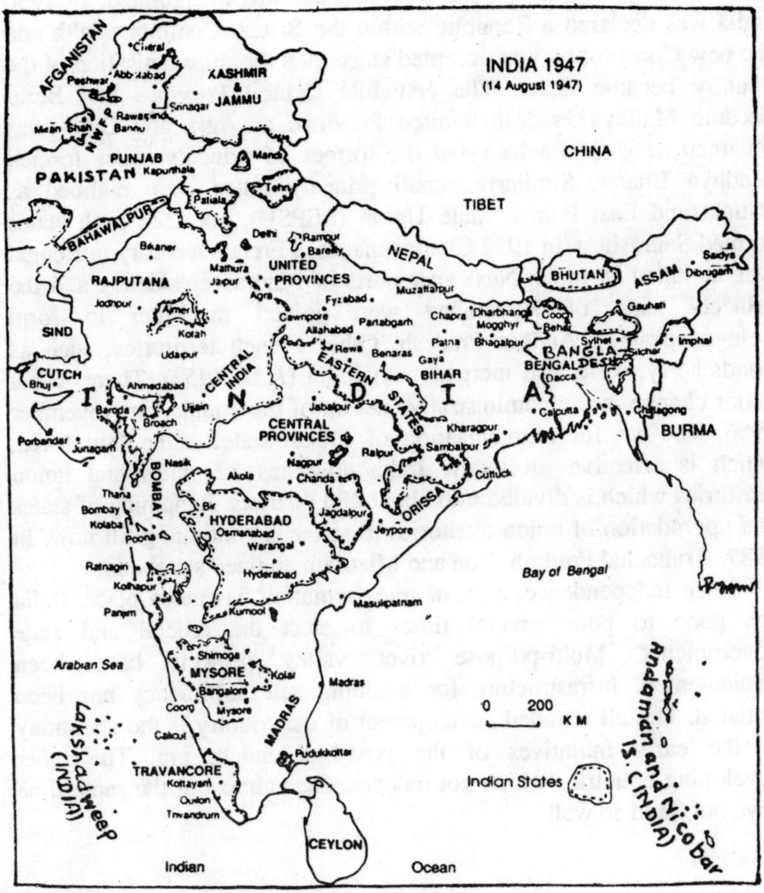

Source: National Atlas of India

MAP 2.2: India: 14th August, 1947

Post-independence Period

After Independence, the process of re-organisation of states continued in India. Hyderabad in the south was taken over by the Indian forces in 1948. Next year United State of Rajasthan was inaugurated. In 1950 India was declared a Republic within the British Commonwealth and the new Constitution was accepted since then the re-organisation of the country became faster. The erstwhile Central Province and Berar became Madhya Pradesh, United Province of Agra and Oudh was renamed as Uttar Pradesh and the former 25 princely states formed Madhya Bharat. Similarly, small princely states were clubbed as Patiala and East Punjab State Union (PEPSU), and 222 such states formed Saurashtra. In 1952 Chandannagar, a French territory in Bengal was included in India. Next year ports of Madras presidency and the princely state of Hyderabad were joined in order to form Telugu speaking Andhra Pradesh. Other French territories, such as Pondicherry, were also merged with India (A.D. 1952). There was a major change in the administrative set up of the country in November 1956. The act for reorganisation of Indian states came into force, which is effective till today. India now has 32 states and union territories which is divided into about 480 districts. Renaming of states and upgradation of union territories to states is continuing till now. In 1987, Arunachal Pradesh, Goa and Mizoram attained statehood.

Since Independence, a lot of transformation has taken place. India has gone to polls several times to elect the federal and state governments. Multi-purpose river valley projects have been implemented. Infrastructure for attaining self sufficiency has been initiated. The all rounded development of the country is the testimony of the early initiatives of the post-Independent era. The other developing countries which got independence almost at the same time have not fared so well.

REFERENCES

Basham, A.L. (ed.), (1975), *A Cultural History of India*, Oxford.

Chatterjee, S.P. (1976), *Geography and Culture*, Sathya Sai Baba Institute, Bangalore.

Chatterjee, S.P. (1982), Evolution of Political History of India as Influenced by geographical factors, *Geographical Review of India*, Vol. 44, No. 1, pp. 1-18.

Nag, P. (1978), The Indian Ocean, India and Africa: Historical and geographical perspectives, in Satish Chandra (ed.), *The Indian Ocean : Explorations in History, Commerce & Politics*, Sage Publications, New Delhi, pp. 151-73.

Panikkar, K.M. (1955), *Geographical Factors in Indian History*, Bharatiya Vidya Bhawan, Bombay.

Singh, R.L. (1955), Evolution of settlements in the Middle Ganga Valley, *National Geographical Journal of India*, Vol. 1, No. 2, pp. 69-114.

Sopher, David E. (1980), *An Exploration of India: Geographical Perspectives on Society and Culture*, Cornell University Press, Ithaca.

Swatzburz, J.E. (1978), *A Historical Atlas of South Asia*, University of Chicago Press, Chicago.

Thapar, Romila (1977), A History of India, Penguin Books, in 2 vols.

Tamaşkar, B.G. (1985), *Contributions to Historical Geography of India*, Inter-India, New Delhi.

Noble, A.G. & A.K. Dutt (eds.) (1982), *India: Cultural Patterns and Processes*, Westview Press, Boulder (Co).

3

GEOLOGY AND STRUCTURE

The most salient fact with regard to both the physical and geological features of India is that it is composed of three distinct units or earth features, which are as unlike in their physical as in their geological characters. The first two of these three divisions of India have a fundamental basis, and the distinctive characters of each which were apparent from a very early period of its geological history. Since then each area has pursued its own career independently. These three divisions are :

(a) the triangular plateau of the Peninsula (i.e. the Deccan south of Vindhyans), with the island of Sri Lanka.

(b) the mountainous region of the Himalayas which borders India to the west, north and east, including the countries of Afghanistan, Baluchistan, and the hill tracts of Myanmar known as the extra Peninsula.

(c) the great Indo-Gangetic Plain of Punjab and Bengal, separating the two former areas and extending from the valley of the Indus in Sind to that of the Brahmaputra in Assam (Map 3.1).

The Southern Peninsula

The Peninsular India has not undergone marine submergence since Archaen times. The Archaen era represented by granites and schists is of the oldest basement complex. During this era highly metamorphosed and folded sedimentary and ignious rocks occurred in the Dharwad district of Karnataka state. Rocks of the same age also outcrop in the Aravalli mountains in Rajasthan and in the northern part of the Eastern Ghats. The Aravalli mountains are the one of the oldest geosyncline of the world. Quartzites, conglomerates and grits are the ridge makers.

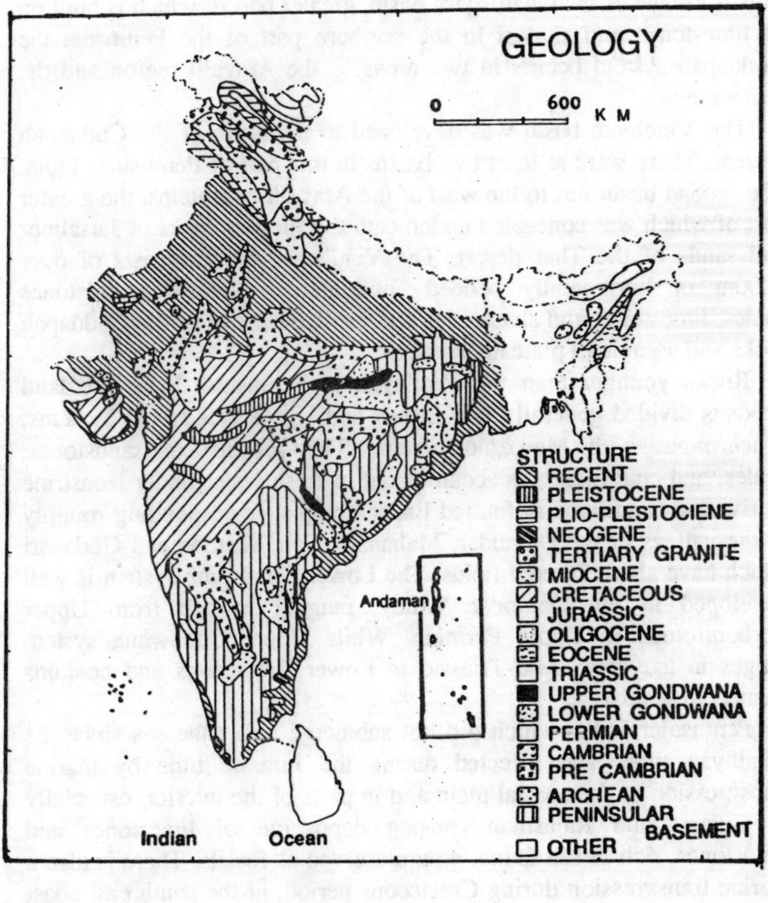

MAP 3.1: Geology

Overlying Dharwar and Aravalli systems rest the rocks of the Purana Group — Cuddapah and Vindhyan systems, unfossiliferous ancient formations, former occurring mainly in Andhra Pradesh and latter in the Vindhyan range, Madhya Pradesh. The second largest basin of the Cuddapah system is found in the upper Mahanadi valley. This is known as the Chattisgarh basin, greater part of which is built up of limestones and shales. In the northern part of the Peninsula the Cuddapah system occurs in two areas — the Aravalli region and the Son valley.

The Vindhyan basin was developed to the north of the Cuddapah system. There were at least two basins in this part of Peninsular India. The second basin lies to the west of the Aravalli mountains, the greater part of which was concealed underneath the Jurassic rocks of Jaisalmer and sands of the Thar desert. The Vindhyan basin consists of over 4000m of horizontally bedded fluviomarine deposits, sandstones shales, limestones and slates over the levelled surface of the Cuddapah rocks and assumes a plateau character.

Rocks younger than the Vindhyan in Peninsular India are land deposits divided generally into Lower and Upper Gondwana Systems, synchronous with Mesozoic system. They consist of sandstones, shales, and coal measures accumulated in fresh water under lacustrine or riverine conditions in faulted linear troughs, corresponding roughly to the valleys of the Damodar, Mahanadi, Son, Wardha and Godavari which have about 70 coal-fields. The Lower Gondwana system is well developed in the Damodar Valley, ranging in age from Upper Carboniferous to Upper Permian. While Upper Gondwana system ranges in age from Mid-Triassic to Lower Cretaceous and contains some coal seams.

Peninsular India, which did not submerge under the sea since the Vindhyan time, was affected during the Jurassic time by marine transgression in the coastal plain and in parts of the interior, especially in Gujarat and Rajasthan causing deposition of limestones and sandstones, rich in sea fauna, now preserved as fossils. There is also a marine transgression during Cretaceous period, in the south east coast of the Peninsula and in Narmada Valley. Cretaceous rocks are mainly composed of limestones, sandstones and shales. During the close of the Cretaceous time Peninsular India experience a violent volcanic out burst as a result of which huge sheets of lava some 3,000 thick, were poured out from the interior mainly through linear fissures, which spread over vast areas of north Deccan, Madhya Pradesh and Gujarat, covering about 520,000 km² of area.

This area possesses the simplest geological structure consisting of flat reposing sheet of lava 650 to 3,000m high, completely burying the ancient geography of this part of the country. For millions of years the lava plateau was subjected to mechanical disintegration and chemical weathering and as a result, flat-topped valleys and plains were formed, which characterise the present day topography of the Maharashtra and Malwa plateau. It may further be summarised that outpouring of so much basaltic magma over the surface might have caused the deepening of the Himalayan foredeep to mountain isostatic balance.

The tertiary deposits in Peninsular India are extremely limited in distribution. The west flowing rivers of the Peninsula like the Narmada and Tapti contain older alluvium of the Pleistocene age. The alluvial deposits of the Narmada Valley are composed of clays with numerous beds of coarse sands and gravel and conglomerates and contain many mamalian bones. The west flowing Tapti also contains older alluvium. The east flowing peninsular rivers, such as the Godavari, Krishna, and Cauvery, have built up terraces of older alluvium in their valleys.

On the top of hills and plateaus of the Deccans and North Central Highlands occur 15 to 60 km thick clay rock of red colour, which belong to Pleistocene Age. On weathering, the soft low land laterite hardens considerably. Laterite also caps Gondwana coal bearing rocks in Ranigunj.

The Northern Mountain

Pre-cambrian metamorphic rocks form the bulk of the Himalaya. These rocks represents the frontal part of the Gondwana which pushed northwards, uplifting the Himalayas as it pressed against the Asian land masses. It is only in the Soiti basin and in a few other localities that large outcrops of marine sediments from Palaezoic and Mesozoic times can be seen. Plutonic rocks, such as granites and granodisrites of Pre-Miocene age cover extensive areas in the northern part of Jammu and Kashmir. Many of the high peaks such as Makalu, Manalsu and Nanga Parbat are formed of gneisses and schists.

The structure of the outer or sub-Himalayan ranges is a generally of great simplicity, they are made up of a series of broad anticlines and synclines of the normal type, a modification of the Jura type of mountain structure. These outer ranges, dissected into a series of escarpments and disslopes, are separated by narrow, longitudinal tectonic valleys or depressions called Duns. Many of the ranges of the outer Himalayas and several of the middle Himalayas as well, are of the orthoclinal type of structure. In the Nappe zone of the Himalayas, in the Simla-Kumaon area, a system of over folds of the recumbent type, served by reserved faults that have passed into thrust planes, along which large slices of the mountains have moved bodily southward.

The structure of the passes of the inner Himalayas in the western part have a comparatively simple type of mountain tectonics. The great sedimentary basins of Hazara and Kashmir lying between the crystalline axis and the zone of the great thrusts (the nappe zone) show a system of normal open anticlines and synclines. The Central zone, however, is not wholly crystalline. Parts of it are capped by unmetamorphosed Palaeo-Mesozoic members of the Tibetan zone of geosynclinical facies, tying to its north.

The broad features of the Himalayas are solely due to movements of uplift, the characteristic scenery of the mountains, the serried lines of range behind range, separated by deep defiles and valleys, the bewildering number of water-sheds, peaks and passages and other rugged features, which give to the mountains their characteristic relief and outline, are the work of the eroding agents, playing on rocks of different structures and varying hardness.

The Great Plains

The third division of India, the great alluvial plains of the Indus-Ganges-Brahmaputra, is least interesting part of India with respect to geological studies. The Great Plain is composed of sand, silt and clay with occasional gravel beds of Pleistocene to Holocene age. Originally, a synclinal depression which begin to form in the Upper Eocene, attained its maximum development during the third uplift of the Himalaya. Since then it has gradually been infilled with sediments.

The exact depth of alluvium has not been ascertained, but it varies from less than 1000 to over 2000 m. Underlying the alluvium are unconsolidated Siwalite and older Tertiary sediments of the Himalayan piedmont and and below these are more consolidated older formations, such as the Gondwanas and the Cretaceous. The depth is greater in the northern than in the southern sector. The depth of the alluvium considered to be about 4600m near its northern limit. The depth is at a maximum between Delhi and the Rajmahal and Assam. The floor of the Gangetic trough is not an even plane, but is corrugated by inequalities and buried ridges. The southern limit of the trough shows no structural peculiarities of features of any importance.

The geology and structure of India has influenced the physiography of the country. Because every geological age has its own physiography, the present surface features of India are the outcome, in a great measure by its geological history.

REFERENCES

Chatterjee, S.P. (1986), *Economic Geography of Asia*, Allied Book Agency, Calcutta.
Singh, R.L. (ed) (1971), *India — A Regional Geography* NGSI Publication, New Delhi.
Wadia, D.N. (1960), *Geology of India*, Mac Millan and Co., London.

4

PHYSIOGRAPHY AND DRAINAGE

India, the seventh largest country in the world, covers an area of 3,287,782 km^2 in the South Asia, with a land frontier of 15,200 km and a coast line of 5,700 km. Physiographically, India, can be divided into five divisions i.e., (a) mountain region of the north; (b) triangular plateau of the peninsula; (c) great Indo-Gangetic plain in between; (d) coastal plains; (e) islands (Table 4.1 and Map 4.1).

TABLE 4.1: Altitude zones

States/Union Territories	Plains below 300 m	Plateau 300-1000 m	Hills 300-2000 m	Mount-ains 2000 m
Andhra Pradesh	48.67	39.01	13.32	-
Arunachal Pradesh	-	-	29.23	70.17
Assam	80.61	-	19.39	-
Bihar	63.46	25.59	10.10	-
Gujarat	85.45	9.80	4.75	-
Haryana & Chandigarh	98.95	-	1.05	-
Himachal Pradesh	-	2.42	39.95	63.63
Jammu & Kashmir	2.77	40.50	6.20	50.53
Karnataka	8.60	76.14	15.26	-
Kerala	64.94	25.09	10.42	-
Madhya Pradesh	14.12	65.38	20.50	-
Maharashtra & Goa & Dadra H.H.	21.09	48.71	30.20	-
Manipur	11.18	35.78	53.04	-
Meghalaya	24.68	18.00	57.32	-
Mizoram	7.11	72.47	19.92	-
Nagaland	15.43	14.52	47.36	22.69
Orissa	60.64	29.68	9.68	--
Punjab	99.07	-	0.93	-
Rajasthan	60.05	27.54	12.41	-
Sikkim	-	13.70	27.40	58.90
Tamil Nadu & Pondichery	64.76	14.94	16.62	3.68
Tripura	67.07	18.61	14.32	-
Uttar Pradesh & Delhi	69.95	9.88	1.85	17.32
West Bengal	85.25	11.34	1.14	2.27
Andaman & Nicobar Island	59.83	-	40.17	-
Total	43.66	33.24	14.80	8.30

Source: S.P. Chatterjee (1986).

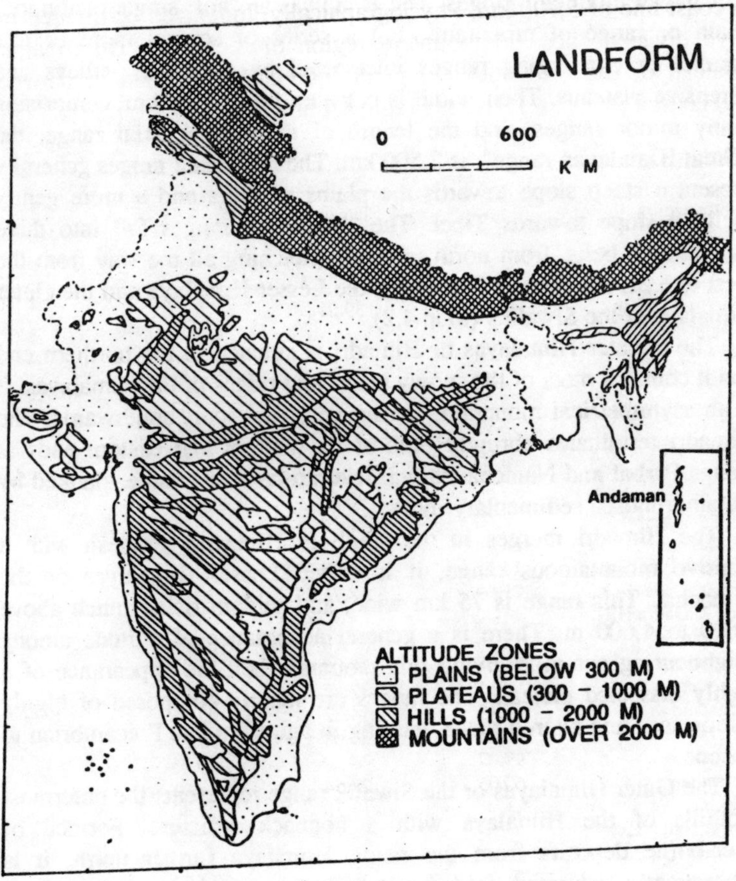

MAP 4.1: Landform

The Northern Mountain Range

High relief, snow-capped summits, deeply dissected topography, antecedent drainage, complex geological structure and rich temperate floras in the sub-tropical latitudes give a distinct character to the Himalayan mountain ranges. The Himalyas are not single continuous chain or range of mountains, but a series of several more or less parallel or converging ranges intersected by enormous valleys and extensive plateaus. Their width is between 160 to 400 km, comprising many minor ranges, and the length of the central axial range, the "Great Himalayan range", is 2,500 km. The individual ranges generally present a steep slope towards the plains of India and a more gently inclined slope towards Tibet. The Himalayan ranges fall into three longitudinal belts, from north to south stretching all the way from the west to east; the Greater Himalaya, the Lesser Himalaya and the Outer Himalaya or the Siwaliks (Map 4.2).

The Greater Himalayas or Himadri is situated in the northern end and it contains most of the snowy peaks and ridges of the Himalayas. It is an asymmetrical mountain, having very few spurs southwards. The Himadri terminates abruptly both at its western and eastern ends at Nanga Parbat and Namcha Barwa. It has a core of granite, flanked by metamorphosed sedimentary rocks.

The Himadri merges in the south in Himachal Pradesh with a massive mountainous range, it is called Lesser Himalayas or the Himachal. This range is 75 km wide, and seldom rising much above 3,600 to 4,600 m. There is a general accordance of altitude among neighbouring summits giving this mountain belt the appearance of a highly dissected plateau. The ranges are mainly composed of highly compressed and altered rocks, varying in age from late Precambrian to Eocene.

The Outer Himalayas or the Siwalik range represents the outermost foothills of the Himalaya with a hogback structure. Formed of river-borne deposits from the rising Himalaya further north, it is subsequently indurated, folded and faulted by earth movement, from the Middle Miocene to the Lower Pliestocene ages. Fault scarps, anticlinal crests and synclinal valleys with an average of elevation of 600 m characterise the Siwalik range. It is bordered at the top by scarps and descends northwards to flat floored structural valleys called *dun* which are intensively cultivated and densely populated. The inner portion of the Siwaliks comprises a series of parallel ranges and structural valleys rising to a maximum height of 1,500 m.

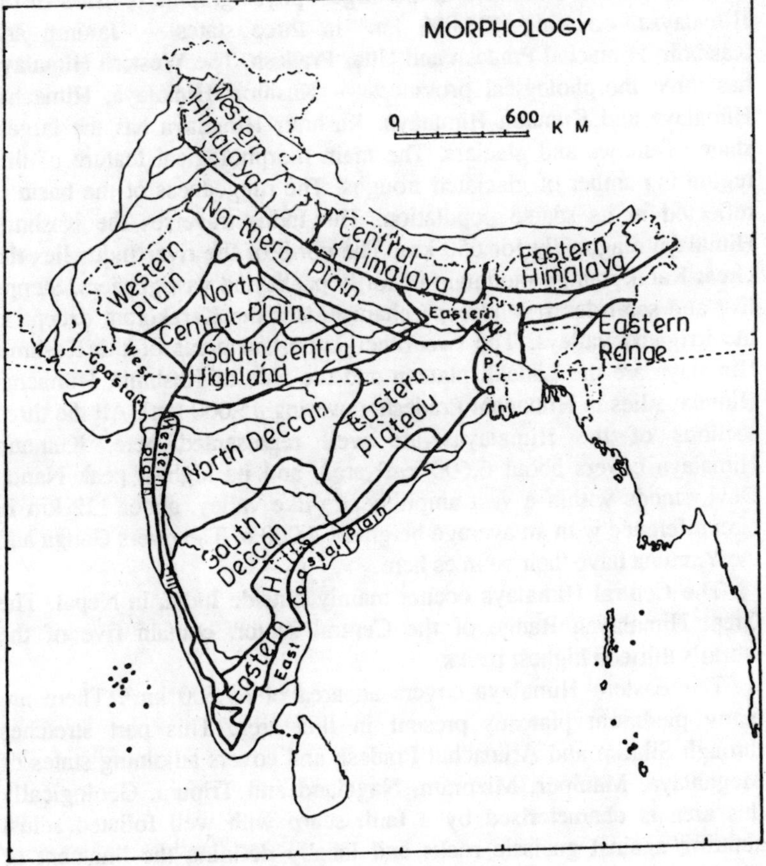

MAP 4.2: Morphology

For geographical purpose, the long alignment of the Himalayan system has been divided into three sections: the Western Himalayas from the Indus to the Kali (880 km long); the Central Himalayas from the Kali to the Tista (800 km long); and the Eastern Himalayas, from the Tista to the Brahmaputra (725 km long).

The Western Himalaya is the largest physiographic division of the Himalayas, covering 433,000 km^2 in three states — Jammu and Kashmir, Himachal Pradesh and Uttar Pradesh. The Western Himalaya has three morphological provinces — Kashmir Himalaya, Himachal Himalaya and Kumaun Himalaya. Kashmir Himalaya has the largest share of snows and glaciers. The main morphological feature of this region is number of glaciated troughs. The ruggedness of the basin is reflected in its sparse population. The Indus traverses the Kashmir Himalaya diagonally for 650 km. The north of the river Indus lies the Great Karakoram, a mountain of loftly peaks and vast glaciers. Steppe like and semi-desert landscape characterises the Karakoram except in the irrigated valleys. The two other important areas of the Kashmir Himalaya are the Ladakh plateau and the vale of Kashmir. Himachal Himalaya lies in Himachal Pradesh, covering 45,000 km^2. All the three sections of the Himalayas are well represented here. Kumaun Himalaya covers about 6,600 km^2 area, and its highest peak Nanda Devi, stands within a vast amphitheatre-like valley, about 112 km in circumference with an average height of 7,000m. The rivers Ganga and the Yamuna have their sources here.

The Central Himalaya occurs mainly outside India, in Nepal. The Great Himalayan Range of the Central sector, contain five of the world's thirteen highest peaks.

The Eastern Himalaya covers an area of 67,500 km^2. There are many piedmont plateaus present in this area. This part stretches through Sikkim and Arunachal Pradesh and covers adjoining states of Meghalaya, Manipur, Mizoram, Nagaland and Tripura. Geologically this area is characterised by a fault scarp with well foliated schist abutting against gneissic rocks and locally defining the linement of artesian spring. The Archaen gneissic shield of Peninsular India continues from the Chota Nagpur plateau of Bihar to Meghalaya after an interruption by the Ganga-Brahmaputra alluvial plains in West Bengal and Bangladesh. This shield occupies more than half of Meghalaya. The Archaen rocks have also been subjected to peneplanation during Dharwarian times and much later in the pre-Cretaceous.

Although the Himalayan region occupies only a little more than 10 per cent of the total area of India, it plays an important role in the physical geography of the country, both as a climatic barriers and as a source of rivers and for their transported materials. The trough of the Great Plains is infilled with the detritus borne down from the Himalaya ranges.

The Southern Peninsular

The plateaux occupy the greater part of Central India and the Deccan, with typical landscapes of extensive flat or rolling plains at altitudes from 300m to 1,000m. In Central India, the Malwa and Vindhya plateaux abut on the Vindhyan mountain range on the south and decline gently northwards. In the northcentral Indian tableland plateau surfaces occur in east Rajasthan and Madhya Pradesh which are called *pathar* or rocky country. The Maharashtra plateau is formed entirely of basalt. In contrast to this plateau, those of Karnataka and Telangana are composed of Archaen geniesses.

The hills in India may be remnants of worn-down mountains, such as Aravalli hills (Rajasthan) the Satpura (northern Maharashtra) or the Eastern Ghat; or dissected plateau escarpments such as Vindhya hills or Sahyadri parallel to the west coast.

The Great Plains

The plains of northern India known as the Great Plains of India constitute a plain of aggregation, built up by three major river systems the Indus, the Ganga and the Brahmaputra. Abandoned river courses, ox-bow lakes and long stretches of marshy land behind river-banks characterises the Great Plains. The Ganga plains are divided into several sub-regions. The Ganga-Yamuna *doab* lies between the rivers of those names. North of the Ganga-Yamuna *doab* lies Rohilkhand plain in Uttar Pradesh. This plain merges into the Avadh plain, covering the eastern half of Uttar Pradesh. Further east lie the lower plains of Bihar, divided by the east-west flowing Ganga into two compact blocks. West Bengal contains the final stretch of the Ganga Plains. The mouth of the Ganga has delta formation with numerous distributaries at the seaward margins. In the west, the Ganga plain merges with the Indus plain which in general is at a higher level than the former. The semi-arid landscape of the southern Punjab plain extends southwards over most of this arid land. The groundwater is saline.

The Coastal Plains

The narrow coastal plains fringing Peninsular India is washed by the Bay of Bengal on the east, the Indian Ocean in the south and the Arabian Sea on the west. The eastern coastal plains are much wider than the western. Several deltas lie on the eastern coast but none on the west. While the western coast runs fairly straight from the Kanyakumari to 22°N, the eastern coastline swings in wide curves, changing direction from north to north-east from 16°N onwards. Unlike the west coast, the east is deficient in rainfall and is backed by shifting sand dunes and stretches of saline sand.

The eastern coastal plain extends through three states — Tamil Nadu, Andhra Pradesh and Orissa. The Cauvery delta is the most important sub-region in Tamil Nadu. In Andhra Pradesh, two of the biggest rivers of the Deccan, the Godavari and the Krishna, flow through this region and form deltas in their lower reaches. In Orissa state it has the Mahanadi delta. South of the Mahanadi delta lies the Chilka lake. The plains on the west coast have width between 10 to 25 km. The northern part of the West Coast lies in Gujarat state, called the Gulf of Cambay coast, where it widens to include the Gujarat plain.

Islands

India has a number of islands both in the Bay of Bengal and the Arabian Sea. The Andamans and Nicobar, lie in the Bay of Bengal between 6°45′ and 14°N and 92° to 94°E. There are three islands in the Andaman group — North, Middle and the South. The South Andaman has the highest hills, Mount Harnet (450m). The coast is very much picturesque. The Great Nicobar is the largest of another group of 19 islands. There are two other islands outside the two main groups — the Barren Islands and the Narcondam Island the first of which was the only volcano within Indian territories though now extinct. Laccadive, Minicoy and Amindiv in the Arabian Sea are now grouped together as a union territory and called Lakshdweep. Built up by corals the islands have fringing reefs very close to their shore. The unique physiography of India has controlled the origin and courses of the vast rivers of the country. This has helped in the creation of an integrated drainage system.

DRAINAGE

Rivers with their tributary systems, are the main channels of drainage of the land surface. They are the chief agents of land erosion and the main lines for the transport of the products of the waste of the land to the sea as well. The drainage systems of the two regions, peninsular and the Himalayan, having had to accommodate themselves to two very widely divergent types of topography, are necessarily very different in their character (Map 4.3).

Peninsular River System

In the peninsula, the river systems are all of great antiquity and consequently, by the degradation of ages, their channels have approached the last stage of river development. The valleys are broad and shallow, characteristics of the regions where vertical erosion has almost ceased and the lateral erosion of the banks by winds rain and streams is of greater momentum. In consequence of their low gradients, the water has but little momentum except during floods, and therefore having a low carrying capacity. In normal seasons they are only depositing agents, precipitating silt in parts of the basins, alluvial banks, estuariene, flats etc., while the streams flow in easy, shallow and meandering valleys.

One notable feature in the drainage system of the peninsula is the easterly trend of its main channels. The Western Ghats, situated close to the western border of the peninsula is the major watershed. The three most important east flowing rivers are the Godavari, Krishna and Cauvery. They contribute greatly to the agricultural prosperity of the Deccan by having built up large deltas, and also by supplying water for irrigation along their courses, wherever dams have been built. Different parts of the catchment areas contribute to the flow of the river various times in the year because of the arrival of the rainly season at different dates within the basins.

The three principal west flowing rivers of central and southern India are the Narmada, Tapti and Sabarmati. Normally these rivers are free from flooding but occasional heavy floods occur.

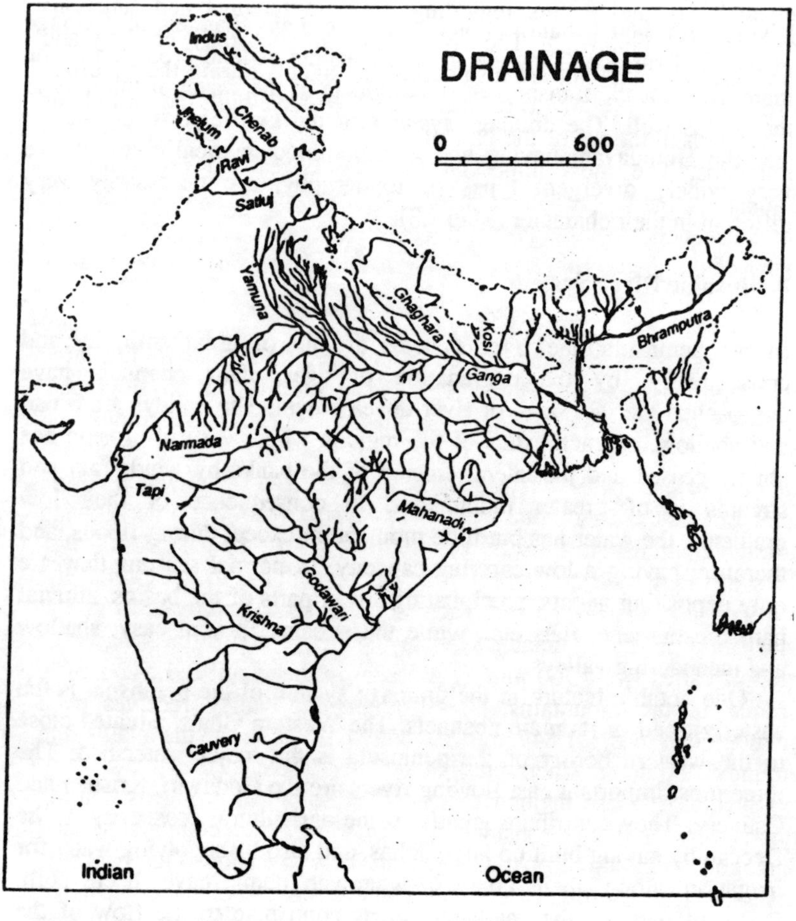

MAP 4.3: Drainage

Himalayan River System

In the Himalayan region the drainage system is comparatively of recent development, and differs regarding its main features and functions from that of peninsula. The rivers here are not only eroding and transporting agents but are also depositing agents during their journey across the plains finally to the sea. Thus they have built the vast plains of northern India out of the part of the silt they have carried from the mountains. The most important fact to be realised regarding the drainage is that it is not in a consequent drainage, i.e. its formation was not consequent upon the physical features, or the relief. In other words, many of the great Himalayan rivers are older than the mountains they traverses.

The Himalayan rivers, fed by melting snows at higher altitudes and by rains almost throughout their courses. This is due to the large catchment basins of the Himalayan rivers span different physiographic and climatic regions. Further, the master rivers, Ganga, Indus and Brahmaputra, have different hydrological conditions and problems in the various sections of their courses. In the Ganga basin, which covers some 861,404 km^2 the total length of streams of the first order flowing throughout Indian territory is 7,552 km. Average annual rainfall over the basin ranges from 1,652 mm at Calcutta (delta) to 980 mm at the confluence of the Ganga and Yamuna (Allahabad) and to 2,327 mm at Mussoorie in the Kumaun Himalaya. In the Indus basin average rainfall varies from 960 mm at Ambala on the plain to 1,542 mm at Shimla in the mountains. The Brahmaputra valley receives even greater quantities : 1,847 mm annually at Tezpur in the plains and around 6,000 mm at Cherrapunji in the hills (Table 4.2).

Several of these major rivers pose hydrological problems. During floods they widen considerably, flooding the countryside and damaging standing crops. Ganga regularly overflows its banks at several places. Shifting of the river courses poses another problem, the Kosi being particularly troublesome. The Brahmaputra valley is subject to more severe flooding than the Ganga basin and serious damage is caused to crops and settlements almost every year. No attempts has so far been made to control this turbulent rivers.

The greater part of the lower course of the Indus now lies in Pakistan. The water of Sutlej, a tributary of Indus, is being utilised for power generation and irrigation in the arid lands of Punjab and Rajasthan.

TABLE 4.2: Catchment area and annual yield of water in the river basins of India.

Rivers	Area in km (%)	Annual yield of water in million m³ (%)	Rate of flow (m³/km²)	Storage capacity million m³
Major rivers				
Ganga	861,404 (26.2)*	468,700 (25.2)	442,170	33,467
Indus	321,284 (9.8)*	79,500 (4.3)	247,441	14,419
Godavari	312,812 (9.5)	118,000 (6.4)	377,223	14,859
Krishna	258,948 (7.9)	62,800 (3.4)	243,403	29,860
Brahmaputra	258,008 (7.8)*	627,000 (33.77)	1,081,034	142
Mahanadi	141,589 (4.3)	66,640 (3.6)	470,658	7926
Narmada	98,795 (3.0)	54,600 (2.9)	970,658	2,550
Cauvery	87,900 (2.7)	20,950 (1.1)	237,770	5,428
Tapi	65,145 (2.0)	17,982 (1.1)	267,770	8,140
Penner	55,213 (1.2)	3,238 (p.2)	58,646	1,978
Brahmani	39,033 (1.2)	18,310 (DO)	202,701	3,953
Mali	34,481 (1.1)	11,800 (0.6)	388,681	4,140
Subarnarekha	19,296 (1.1)	7,940 (0.4)	411,484	283
Sabarmati	21,895 (0.7)	3,800 (0.2)	173,556	1,017
Medium and minor	711,833 (21.7)	296,840 (16.0)	417,008	18,452
Total (India)	3,287,782 (100.0)	1,858,100 (100.00)	565,153	146,826

* Area within India.
Source : S.P. Das Gupta, 1989.

Lakes play very little part in the drainage system of India. Even in the mountainous regions of the Himalayas, lakes of notable size are very few. The small fresh-water lakes of Kashmir are regarded as mere inundated hollows in the alluvium of Jhelum. There are few salt lakes in the peninsular region which are however part of the drainage system of the country.

REFERENCES

Chatterjee, S.P. (1973), *India, in World Atlas of Agriculture*, Asia & Oceania Volumes, International Association of Agricultural economists. Novava, pp 186-200.

Chatterjee, S.P. (1974), Himalayan Mountain Ranges, in *New Encyclopaedia Britannica*, Chicago, Vol. 8, pp. 882-7.

Chatterjee, S.P. (1984), *Economic Geography of Asia*, Allied Book Agency, Calcutta.

Chattérjee, S.P. (1975), Physiography, in *Gazetteer of India*, Chapter I.

Chhibber, H.L. (1948), The drainage pattern observed in India and adjacent Countries, *National Geographical Journal of India*, Vol 2 (1), pp. 2-5.

Das Gupta, S.P. (1989), India's water resource potential, in S.P. Das Gupta (ed), *Man and Ecology*, School of Fundamental Research, Calcutta.

Gansser, A (1964). *A Geology of Himalayas*, Inter Science Publisher, London.

Law, B.C. (1968), *Mountains and Rivers of India*, 21st International Geographical Congress Publication, Calcutta.

Singh, Jasbir (1984), Water resources in India : A geographical perspective, *Geographical Review of India*, Vol. 46, pp. 79-87.

Singh, R.L. (1971), *India — A Regional Geography*, NGSI Publication, Varanasi.

5

CLIMATE

India has a typical monsoon climate with two seasonal winds: the north-east (winter) monsoon and the south-west (summer) monsoon. India has great variation in rainfall and temperature from one part of the country to other part. The mean annual temperature of India ranges from below 20°C to above 27.5°C. Considering the year round temperature, the country can be divided into five temperature zones. The 20°C isothermal line limits the southern extension of the cool Himalayan region where high altitude has a direct influence on lowering temperature.

Seasonal temperature of India get modified considerably by the south-west monsoon winds; hence the normal classification of the seasons of a country into winter, spring, summer and autumn cannot be exactly applied here. Winter generally lasts for three months, December to February. The spring season is almost non-existent except in some local areas. Its place is substituted by the early summer or pre-monsoon season which lasts for three months, March to May which used to have occasional showers. Then follows the rainly season with fairly high temperature, somewhat modified by heavy rains which last for four months, June to September. The last two months of the year have fine weather, this is the autumn season or as it is called, the retreating monsoon period (Table 5.1).

Temperature

By far the largest part of India lying between 15° and 30°N has more or less uniform temperature, enclosed by 29°C and 27°.5°C isothermal lines. The determination of the thermal conditions as expressed by the mean daily maximum and minimum temperatures is of importance in evaluating cropping pattern in an agricultural country like India. The range between the mean of daily maximum 33.6°C and the mean of daily minimum 18.2°C is the highest in Bikaner — 15.4°C or 16°C.

Leh, the cold desert high altitude station, has a mean maximum diurnal temperature of 12.3°C but its mean minimum temperature drops to — 1.4°C. In lower altitudes the maximum daily temperature is the second highest at Nagpur, the geographical centre of India, about 2°C less than that of Bikaner. The range of diurnal temperature is about the same in the east and west coasts but the maximum daily temperature on the east coast is higher than that of the west coast.

TABLE 5.1: Rainfall and area

Annual Average Rainfall (mm)	Receiving	Area
	km²	% of total area
Over 3000	110,184	3.36
2,000 - 3,000	83,560	2.54
1,000 - 2,000	1,679,823	51.16
800 - 1,000	380,288	11.58
400 - 800	735,254	22.39
200 - 400	183,896	5.60
Under 200	110,231	3.37
Total	3,283,236	100.00

Source: S.P. Chatterjee (1972).

January is the coldest month in the year when dry continental winds blow over the greater part of India. Temperature increases continuously from the foothills of the Himalaya (10°C) in the north to Kanyakumari (25°C) in the south because of the meridianal flow of atmospheric circulation from north to south. During this month India falls into five temperature regions — very cold, cold, cool, warm, and hot, marked by 10°C, 15°C, 20°C and 25°C isothermal lines respectively. The first three regions run more or less parallel from west to east, the fourth runs longitudinally in and from the coasts and the fifth runs along the west coast in narrow belt.

The hot weather, from March to May, is of continuous increase of temperature and decrease of barometric pressure in northern India, and of continuous decrease of temperature in the southern India. A steady transference of the area of greatest heat occurs northwards. In March the highest day temperature of about 37.7°C occurs in the Deccan; in April the area of the highest day temperature from 37.7°C to 43.3°C lies in the south of Madhya Pradesh and Gujarat. In May the seat of the greatest heat is in northern India and especially in the north western *Marusthali* desert where temperature often reads 48.8°C or above. The area of lowest pressure lies over *Marusthali*.

July is the hottest month of the year for many parts of India, though monsoon rains lower the temperature after each downpour. In October there is an uniform distribution of temperature, around 28°C, practically throughout the country. Spells of abnormal maximum and minimum temperature occur over various parts of the country. Heat waves come between April and June; they are experienced in the north and central India. The cold waves occur mostly in Jantary and February in the same region.

Humidity and Rainfall

The humidity of the air causes discomfort in summer months. The relative humidity is lowest in Punjab and increases towards the sea in all directions. In the monsoon season the air is nearly saturated. During the cold months, the isohyet ranges between 70 to 80 per cent, whereas in the hot month it rises to 90 per cent.

Rainfall is the most significant climatic element in India and is unevenly distributed in the country ranging from less than 150 mm at Leh to 6,000 mm at Mahabaleswar and Cherrapunji. On the basis of annual rainfall India can be divided into eight rainfall zones: (a) very wet, (b) wet, (c) moderately wet, (d) sub wet, (e) sub-dry, (f) dry, (g) hot desert, (h) cold desert.

The northeast monsoon begins in the later part of December and gets fully set by the beginning of January in the land and sea areas of the country. Then the temperature is lowest in the central and western Asia. There is belt of high pressure with anticyclonic conditions stretching from the west Mediterranean to central Asia. Clear sky, fine weather, low humidity, large diurnal range of temperature and northerly winds are the usual features of the weather in India during winter months, only interrupted at intervals by weather disturbances which originate in the Mediterranean Sea and pass eastwards to northern India. The precipitation accompanying them is less. During this period of the year, rainfall or snowfall is greatest in the north west and decreases towards south and east. The retreating north east monsoon continues to give occasional rains in December on the east coast of southern Deccan in association with cyclonic storm and depression originating from the Bay of Bengal.

Western disturbances continue to enter India in March and bring rains. In their further eastward passage they give rise to thunderstorms which are often accompanied by hail in the Great Plains. As hot season advances, thundershowers occurs more frequently and rainfall becomes heavier. Away from the coast, in the inland plains and plateaus, the rainfall in the hot season often preceded by cyclonic disturbances.

Towards the end of May the air circulation over India becomes more and more vigorous until south east trade winds from the south of the equator are induced northwards into the Arabian Sea and the Bay of Bengal and finally becomes part of air circulation of the country. This relatively cool and humid current known as southwest monsoon, bursts around the Malabar Coast in southern India during the early days of June. It is the great rain bearing wind for most of India.

India's rainy season comes with the southwest monsoon during June to September, except in southeastern Tamil Nadu where the main rainy season occurs from October to December (Table 5.2). The southwest monsoon enters India into two main currents from the first week of June to the first week of July. It is in July, when the rainy season is well established, that the entire country becomes overrun by humid monsoon air. The monsoon strikes the west coast from the Arabian Sea and brings heavy rains to the coastal plains and hills, and eventually reaches Gujarat on the central parts of the country. The current originating from Bay of Bengal after crossing the West Bengal coast, is directed west wards up in the Ganga plain by the deflecting action of the mountains to the west and north of West Bengal. In June, rainfall is high along the west coast and the western scrap of Sahyadri. Cherrapunji in Meghalaya receives the maximum rainfall.

Monsoon rainfall ceases in eastern Uttar Pradesh early in October. During the subsequent withdrawal of the Bay of Bengal branch of the monsoon from northern India, those moist winds, continue to give rains in northeastern India until their final retreat by the middle of the month. About this time the Arabian Sea branch also withdraws from the central parts of the country. The amount of rainfall decreases from the coast towards the interior. The coastal district of Tamil Nadu receive most of their rainfall in this season.

Climatic Regions

The climatic characteristics of India are very much contrasting. It is not an easy task to divide the country into climatic regions. However, seven climatic regions can be identified (Map 5.1).

TABLE 5.2: Normal rainfall (reference year 1985-86) in millimetres

Regions	Post-Monsoon 1-10-85 to 31-12-85	Winter Monsoon 1-1-86 to 28-2-86	Pre-monsoon 1-3-86 31-5-86	S.W. Monsoon 1-6-86 to 30-6-86	Total
1	2	3	4	5	6
Andaman & Nicobar Island	694	133	594	1584	2960
Arunachal Pradesh	173	102	783	2060	3118
Assam & Meghalaya	149	41	635	1614	2444
Nagaland, Manipur, Mizoram & Tripura	211	39	503	1084	1837
Sub-Himalayan West Bengal Sikkim	185	56	623	1961	2825
Gangetic West Bengal	152	35	159	1061	1407
Orissa	181	35	119	1146	1481
Bihar Plateau	103	50	108	1088	1349
Bihar Plains	73	31	95	1015	1214.
Eastern Uttar Pradesh	66	36	37	884	1023
Plains of Western Uttar Pradesh	47	43	37	790	917
Hills of Western Uttar Pradesh	76	109	162	1166	1513
Haryana, Chandigarh and Delhi	32	41	43	485	601
Punjab	32	57	48	485	622
Himachal Pradesh	96	156	212	973	1437
Jammu & Kashmir	122	219	363	573	1277
Western Rajasthan	14	10	20	291	1612
Eastern Rajasthan	34	15	21	636	706
Western Madhya Pradesh	53	20	24	926	1023
Eastern Madhya Pradesh	75	45	62	1168	1350
Gujarat Region, Daman, Dadra & Nagar Haveli	31	3	12	897	943
Saurashtra, Kutch & Diu	34	3	6	528	571
Konkan & Goa	132	2	59	2706	2899
Madhya Maharashtra	107	5	57	833	1002
Marathwada	80	13	36	709	838
Vidarbha	76	23	44	968	111
Coastal Andhra Pradesh	313	17	80	614	1024
Telengana	111	12	74	757	954
Rayalaseema	218	12	78	382	690
Tamil Nadu and Pondicherry	484	49	138	322	993
Coastal Karnataka	261	2	181	3062	3506
North Interior Karnataka	125	9	92	520	746
South Interior Karnataka	222	12	191	819	1244
Kerala	500	39	594	2139	3272
Lakshadweep Island	329	37	195	946	1507

Source: Indian Meteorological Department.

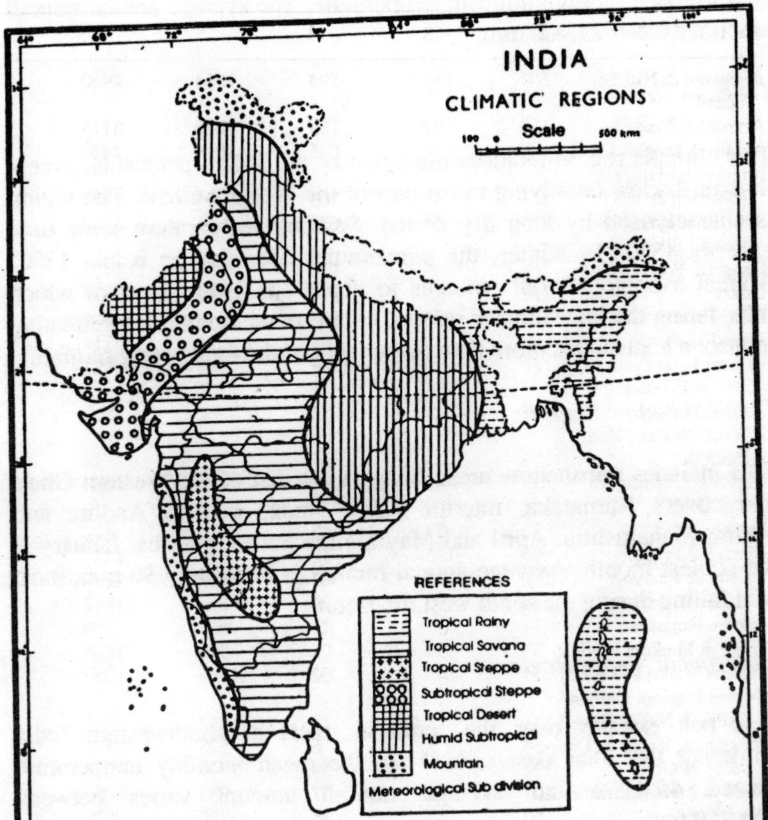

MAP 5.1: Climatic Regions

Tropical Rainy Climate

This belt comprises of Western Ghats, south of Bombay, western coast, Meghalaya, western part of Nagaland and Tripura. This region is characterised by high uniform temperature. The average annual rainfall amount exceed 25,000 mm.

Tropical Savana

This climatic region includes most part of the Indian peninsula, except the rainshadow area lying to the east of the Western Ghats. The region is characterised by long dry period. Summer temperature some time exceeds 45°C. In winter, the temperature does not go below 18°C. Annual average rainfall amounts to about 1,000 mm most of which falls during the late summer season. In the southeastern coast retreating monsoon contributes more than 50 per cent of the total yearly rainfall.

Tropical Steppe Region

This includes rainshadow areas lying to the east of the Western Ghats and covers Karnataka, interior Tamil Nadu, western Andhra and central Maharashtra. April and May are the hottest months. January is the coolest month. Average annual rainfall is less than 750 mm, most of it falling during the south west monsoon.

Subtropical Steppe Region

This belt extends over the semiarid tracts of southwestern India bordering the Thar desert. In May, the mean monthly temperature exceed 45°C annual average rainfall amount varies between 500-750 mm.

Tropical Desert Region

Western Rajasthan region is characterised by high diurnal and annual ranges of temperature. May and June are the hottest months when the average monthly temperatures rise to over 45°C while in January it comes down to 12°C. Annual rainfall amount is less than 150 mm.

Humid Subtropical Region

This region includes the northern plains of India. Highest summer temperatures rise to over 40°C. Rainfall amount varies widely from east to west. In the eastern part the annual rainfall is over 250 cm while it decreases to less than 650 mm in the extreme west.

Mountain Region

This region includes the entire Himalayan belt, where altitude induces very low temperatures. During the summer time temperatures hardly exceed 17°C and in winter it comes down to -8°C. Rainfall is heavy in the foothill regions of the eastern Himalayas where it amounts over 2,500 mm annually. In the western Himalayas the southern part receives an annual rainfall in less than 100 cm.

Recent Attempts to Study Monsoon

India's primarily rural economic is based on climate, more precisely on the monsoon. The delays and disturbance in the process have wider ramifications. Floods and droughts as the consequences of abnormal climatic conditions, have seriously affected the Indian economy. Their increasing frequency and severity have lead to catastrophies of all sorts. The Department of Science and Technology in association with the India Meteorological Department have developed a computer-based Long Range rainfall forecasting model. This power regression model is to predict monsoon rainfall based on 15 parameters, some of which are global and other one regional in nature. The data of last 37 years (1951 — 1987) were analysed to forecast 1989 monsoon. The results were encouraging. It was observed that whenever more than 70 per cent parameters indicated favourable signals, the monsoon is on the positive side of the normal.

This model is purely a qualitative decision-making tool where equal weightage was given to all the 15 parameters. But, these parameters do not have linear relationship with monsoon rainfall. In order to circumvent this problem a curvilinear relationship has been determined by fitting the equation of different degrees and the power regression model was developed. Recently, one more parameter has been added, viz. northern hemisphere pressure anomally. Attempts are also being made to develop methodology for quantitative forecast of rainfall.

With the present model, rainfall performance for every week can be predicted fairly accurate. In 1989, there was no "scanty" rainfall situation in any of the 35 meteorological sub-divisions. Even in the 6 sub-divisions which were "deficient" the rainfall in as many as 5 was only slightly below "normal". Only in East Rajasthan sub-division, the rainfall was significantly deficient (40% below normal). During the monsoon season (June, 1-30 September, 1989), roughly 80 per cent of the country's land received normal or excess rainfall. In the remaining area, the rainfall was slightly below normal, except in parts of east Rajasthan, adjoining parts of southwest Uttar Pradesh, Delhi and northwest Madhya Pradesh, which is about 5 per cent of the area of the country. This year the typical drought prone areas received good rainfall. On the other hand, the flood prone areas of river Ganga did not received major floods. Table 5.3 indicates the performance of the monsoon in the earlier years of the decade.

TABLE 5.3: Performance of the monsoon rainfall

Year	Total monsoon rainfall over the country as a whole
1981	100% of normal
1982	86% of normal
1983	113% of normal
1984	96% of normal
1985	93% of normal
1986	87% of normal
1987	81% of normal
1988	119% of normal
1989	101% of normal

Source: Monsoon 1989 : End-of-Season Review, DST/IMD.

The perfection of the above model will surely help in estimating rainfall. This will help in agricultural management of the country.

REFERENCES

Alvi, S.M.A. and P. Koteswaram (1985), Time series analysis of annual rainfall over India, *Mausam*, Vol. 36, No. 4, pp. 479-90.

Das, N. D.S. Desai and N.C. Biswas (1989), Weather-Monsoon season (June-September 1988), *Mausam*, Vol 40, No.3, pp. 351-64.

Mooley, D.A. and J. Shukla (1989), Index of activity of the monsoon trough over India, *Mausan*, Vol. 40, No. 3, pp. 247-58.

Mooley, D.A. (1975), Vagaries of the Indian summer monsoon during the last 10 years. *Vayu Mandal*, Vol. 5, pp. 65-6.

6

SOIL

The soils are either the altered residue of the underlying rocks or the deposition of alluvial debris brought down by the river. On the whole, the principal character of the main soil groups of India are generally correlated to their geological foundation. Traditionally soils of India are classified into : (a) the alluvial soils, (b) black soil or *regur*, (c) red soils, (d) laterites, (e) forest soils, (f) mountain soils, (g) desert soils, (h) saline and alkaline soils, and (i) peaty soils (Map 6.1).

The differentiation of climatic conditions, variation in topographic characters and altitudes and in the growth of natural vegetation are all responsible for differentiating soil types of the country. Usually the northern mountains, the great plains and the peninsular India have distinct types of soil (Table 6.1).

Alluvial Soils

The alluvial soils are water transported and have adequate potash content but are deficient in nitrogen content. The alluvial soils spread in a vast sheet between the river Sutlej on the west to the lower Ganga-Brahmaputra valley in the east, covering 1.5 million km^2 of area. The physical and physico-chemical characteristics of alluvial soil varies considerably. The main features of the soil are derived from the deposition and silt laid by the numerous river system. Generally, the deposit is of silty nature which confers fertility on the soil. The colour varies from light grey to ash-grey and the texture is sandy to silty loam. The alluvial soil is both well drained and poorly drained. In some places it has high textured subsoil while in the other places the subsoil is low textured. These soils have an immature profile in undulating land while flat lands have a more mature profile. The coastal alluvial soils are of tidal origin. These soils when reclaimed are rich in plant nutrients and support a good crop. The deposits are rich in calcium and magnesium and also contain sometimes half decomposed organic matter. The alluvial soils are the most productive of Indian soils, contributing the largest share of the agricultural wealth of the country.

INDIA
SOILS
Scale
REFERENCES
Alluvial Soils
Coastal Alluvium
Desert Soils
Deep Black Soils
Black Soils
Mixed Red & Black Soils
Red Soils
Red & Yellow Soils
Lateritic Soils
Brown Forest Soils
Grey & Brown Soils
Hills & Tarai Soils
Forest Soils
Mountain Soils
Snow & Skeletal Soils
Peaty Soils

MAP 6.1: Soils

TABLE 6.1: Major soil types

	Area (million ha)	%
Alluvial Soils	101.2	
Deltaic alluvium, occasionally saline	17.0	
Alluvial soils, highly calcareous	8.9	
Coastal alluvium	8.5	
Alluvial soils affected by salinity and alkalinity	6.9	
Total alluvial soils	142.5	43.4
Red soils	3.1	
Red gravelly soils	1.6	
Red and Yellow soils	17.8	
Mixed red and black soils	10.5	
Total Red Soils	61.0	18.6
Medium black soils	18.6	
Deep black soils	6.9	
Shallow black soils	4.9	
Black soils affected by salinity and alkalinity	6.9	
Black soils, undifferentiated	12.5	
Total black soils	49.8	15.2
Laterites	10.1	
Laterites and lateritic soils	2.1	
Total lateritic soils	12.2	3.7
Desert Soils	14.6	
Mountain meadow soils	11.7	
Mountain soils, undifferentiated	6.5	
Piedmont and terai soils	5.7	
Forest soils, laterised	6.5	
Podzolic soils	3.6	
Hill soils	2.4	
Grey and brown soils	3.6	
Brown soils under deciduous forests	1.6	
Skeletal soils	2.4	
Peat, muds and humose soils	0.2	
Total other soils	58.8	17.9
Area under snow and glaciers	4.0	1.2
Total reported area	328.3	100.0

Note: The reported area is 99.88 per cent of the country's area.
Source: S.P. Chatterjee (1973 & 1986).

Black Soils

Black soils covers an area of 0.5 million km² in Maharashtra, Western Madhya Pradesh, parts of Karnataka, Andhra Pradesh, Gujarat and Tamil Nadu. It is derived from two types of rocks – the Deccan or Rajmahal trap and ferrogenous geniesses and schists occurring in Tamil Nadu. The colour of the black soils varies from deep black to grey. The colour of the soil is dependent on the colour of the mechanical fractions. In general, the black soils have clay texture, average clay content being 50 per cent and the range being 40-50 per cent. Except in cases where there is stratification, the clay content down the profile is uniform. The structure of these soils is usually cloddish but occasionally friable. The black soils of India are calcereaus neutral to mild alkaline in reaction, high in cation exchange capacity and low in organic matter. In general these soils have high degree of fertility, but some especially in the uplands are having low fertility. In the valley the soils are darker, deeper and richer. They contain a high proportion of calcium and magnesium carbonates and iron. Although the potash content varies these soils are moderately well supplied. However, these are poor in organic matter, nitrogen and phosphorous. The medium and deep black soils are very suitable for cotton cultivation.

Red Soils

The ancient crystalline and metamorphic rocks on weathering have given rise to red soils. These soils are found in large tracts of Tamil Nadu, Karnataka, southern Maharashtra, eastern Andhra Pradesh, Madhya Pradesh, Orissa and Chota Nagpur plateau of Bihar. It is also found in the sedimentaries of the north eastern mountains. The texture of red soils varies from sand to clay, the majority being loams. The colour of these soils is mainly due to ferric oxides occurring as thin coatings on the soil particles while the iron oxide occurs as haematite or anhydrous ferric oxide, the colour is red and when it occurs in the hydrate form as limonite the soil gets a yellow colour. Ordinarily the surface soils are red while the horizon below gets yellowish colour. The soils are either shallow or deep. The characteristics of clay mineral present is kaolinite type. The cation exchange capacity of the soil and the base status are rather low.

In many places black and red soils occur side by side and very close. These soils exhibit the properties corresponding to them. The red soils are found at higher elevations, the black soils are at lower position. Mixed red, yellow and grey soils are all generally sedentary soils formed from the underlying parent rock which is mostly granite. The texture of surface soil varies in the first six inches from stony to sandy and loamy in the pale yellow to light red groups, from loamy to clayey in the light red to deep red groups and in the grey soil groups.

Laterites

Laterites area, end product of decomposition, found in regions of heavy rainfall which promotes leaking of lime and silica. They are yellow-grey coloured on top and red below. Organic matter, nitrogen, available phosphate and calcium are low and the soils have acidic reaction. In some laterite soil the top soils may be deep and gravelly but the laterite proper is immature. The soil is rich in iron and aluminium compounds. There soils are generally red due to the presence of iron oxide. Laterites are wide spread in peninsular India. Due to intensive leaching and low base exchange capacity, typical lateritic soils are generally infertile and of little value for crop production.

Forest Soils

Forest soil occur between 3,000-3,100 m in the coniferous forest. The whole area is covered by dense forests. The ground is covered with forest litter. The soils below the litter is black in colour containing a lot of decomposed organic matter. Below these are grey-brown or greyish red soils. These soils being virgin are very rich in plant nutrients but pH is however low.

Mountain Soils

Mountain soils are found on high peaks at altitudes of 2,100 m to 3,000 m. Mostly it is under forests. These soils are generally shallow in depth. The carbon nitrogen ratio is very wide. They are silty loam to loam in texture and dark brown in colour. The soil reaction is slightly acidic to moderately acidic. The surface drainage is very rapid. The internal drainage is very good.

Desert Soils

A large part of the arid and semi-arid region belonging to Rajasthan, Punjab and Haryana where annual rainfall is less than 50 cm is affected by desert conditions. This part is covered under a mantle of browns which inhabits soil growth. Texturally these soils are coarsely sandy. The pH is on the alkaline side. Over large parts of these areas it is seen that calcium content increases downwards resulting *kankar* bed formation. The calcium content in the subsoil is nearly ten times that of the top soil. The phosphate contents of these soils compare favourably with some normal alluvial soils. The total nitrogen component is low. This deficiency is however, made up to a certain extent by the presence of high available nitrogen in the form of nitrates. Thus phosphates together with nitrates go a long way to make these desert sands fertile for agricultural crops wherever moisture is not deficient. Even though the clay content is low in desert sands, there is a tendency to high dispersion on account of sodium clay formation in the subsoil which may render it less permeable and therefore, more retentive to moisture.

Saline and Alkaline Soils

Saline and alkaline soils occur in the drier parts of Bihar, Uttar Pradesh, Haryana, Punjab, Rajasthan and Maharashtra. During dry periods the salts are deposited on the surface in the form of crust or efflorescence. Imperfect drainage, because of cemented layer and a high percentage of salts, interfere with the growth of most normal crops. It is interesting to note that in most of the places these were normal, and well-drained cultivated soils. Originally the advent of canal system is supplying big volumes of water to an already saturated area. Soils, where water is not on the surface but within capillary rise range, have also been adversely affected and show efflorescence of salts on the surface during dry periods. Texturally, they are sandy to loamy sand. They are deficient in calcium, nitrogen and hence very low water bearing capacity.

Peaty Soils

Peaty soils originate in the humid regions due to the accumulation of large amounts of organic matter in the soils. They are generally submerged under water during the monsoon season. The sõils are black and heavy and highly acidic. Sometimes the soils are highly toxic to plant life as they contain ferrous and aluminium sulphates in considerable amounts. These soils have an accumulation of large quantities of water solvable alkali salts.

Soil-related Issues

There are several issues associated with soils. Since agriculture is the source of livelihood of about 75 per cent of the population, these factors become important for planning and development strategy. By the turn of century, apart from 10 million ha additional land for agriculture, about 40 million ha for production of fuel and wood and 10 million ha for fodder will also be required. Obviously, the marginal lands including areas having poor soil cover will have to be brought under more productive use.

The problems due to soil erosion and land degradation affect large areas of the country (Table 6.2).

TABLE 6.2: Soil erosion and land degradation

Problems	Million ha	(%)
(a) Area subject to wind erosion and aridity	38.74	(11.78)
(b) Area affected by sand dunes	7.00	(2.13)
(c) Water logged area	6.00	(1.82)
(d) Alkali soils	2.50	(0.76)
(e) Saline soil including coastal sandy areas	5.50	(1.67)
(f) Ravines and gullies	3.97	(1.21)
(g) Areas subject to shifting cultivation	4.36	(1.33)
(h) Riverine and torrents	2.73	(0.83)
(i) Total flood prone areas	40.00	(12.16)
(j) Annual average area affected by floods	9.00	(2:74)
(k) Annual average cropped area affected by floods	3.80	(1.15)
(l) Drought Prone areas	260.00	(79.03)

Source: Indian Agriculture in Brief.

Soil erosion occurs mainly due to wind, aridity and sand dunes. There is about 45.74 million ha area affected by this problem. Rajasthan, Madhya Pradesh and Maharashtra each have approximately 15 per cent of this area (Table 6.3). Further, there are additional problems which makes areas degraded. About 25 million ha area is under this category which is due to water logging, alkalinity, salinity including coastal salinity, ravines and gullies, shifting cultivation and due to riverine and torrents. In addition, the use of good soils is restricted due to floods and droughts. Where the floods are regular features, adjustments have been made in agricultural practices. But every year a new area is affected by this problem. Rajasthan has about 39 per cent of the degraded land of the country, followed by Uttar Pradesh (13%). About 17 per cent of all types of problem areas were treated by 1984-85. Rest is yet to be treated. The details about the floods and droughts are discussed in the following chapter.

TABLE 6.3: State-wise distribution of area subjected to soil erosion and land degradation (%)

State	Soil Erosion	Land Degradation
Andhra Pradesh	9.08	1.63
Assam	1.75	1.74
Bihar	3.36	5.09
Gujarat	7.85	5.87
Haryana	1.25	5.72
Himachal Pradesh	1.51	—
Jammu & Kashmir	0.69	0.02
Karnataka	8.69	0.92
Kerala	1.39	0.39
Madhya Pradesh	15.49	2.46
Maharashtra	15.15	1.48
Manipur	0.30	0.80
Meghalaya	0.66	0.59
Nagaland	0.32	0.18
Orissa	3.61	7.17
Punjab	0.80	4.94
Rajasthan	15.72	38.87
Sikkim	0.24	—
Tamil Nadu	2.87	0.40
Tripura	0.13	0.22
Uttar Pradesh	5.62	13.34
West Bengal	0.82	7.27
Union Territories	2.70	0.90
Total	100.00	100.00

Source: Indian Agriculture in Brief.

India has a variety of soils, some of which are extremely good for agricultural production. But, due to climate, environmental and man-made problems, they cannot be wholly exploited. The treatment to reclaim good soil has been slow till now. Due to the teeming millions, the pressure on such areas will continue to increase in the years to come.

REFERENCES

Chatterjee, S.P. (1973), India, in *World Atlas of Agriculture*, Novara.

Chatterjee, S.P. (1986), *Economic Geography of Asia*, Allied Book Agency, Calcutta.

Kayastha, S.L., (1965), Some aspec s of soil erosion and conservation in India, *National Geographical Journal of India, Vol. XI(1), pp. 22-9.*

NATMO, (1978), Atlas of Agricultural Resources of India, *National Atlas & Thematic Mapping Organisation, Calcutta.*

Roychaudhuri, S.P. (1986), Land and Soil, *National Book Trust, New Delhi.*

7

ENVIRONMENT

Recently among the planners, decision-makers, academicians and people at large there has been a lot of concern regarding the environment, its problems, prospects, management and social and economic fallouts. As a result, activists and non-government organisations involved in different aspects of environment have sprung up all over the country. The governments at national and state levels are forced to consider environmental problems which may be of local significance but of national concern. Due to this fact the spatial approach to national and state planning is now considered significant. This is a major shift in the planning policy of the country.

Forest have been considered as the most significant aspect for environment considerations in the country. The majority of the so-called natural environmental problems have been thought to be linked with the depletion of the forests. Following this, flora and fauna have also been studied by the scientist as an essential part of the environmental protection strategy. The environmental issues which have been dealt here have some links with the natural and man-made environment.

Droughts

Excepting a few regions and in specific months, there has been trend of reduction in rainfall as available from the latest meteorological data. This trend has been responsible for creating drought like situation in the large areas of the country recently, though irrigation has been reducing the possible losses due to severe and chronic droughts. Government has sponsored several Drought Prone Area Development projects during the 7th plan. Almost one project in each drought effected district has been initiated (Table 7.1).

TABLE 7.1: Drought prone projects

State	No. of Project
Andhra Pradesh	8
Bihar	5
Gujarat	9
Haryana	1
Jammu & Kashmir	2
Karnataka	11
Madhya Pradesh	6
Maharashtra	12
Orissa	4
Rajasthan	8
Tamil Nadu	6
Uttar Pradesh	16
West Bengal	3
Total	91

Source: Indian Agriculture in Brief.

Drought prone projects are also running in the areas which were never considered as water deficit areas earlier, such as Udhampur, Almora and Garhwal (Map 7.1). The central and state governments spent Rs. 177.50 crores during the 5th Plan (1974-78) which rose to Rs. 337.41 crores in the 6th Plan (1980-85) for the development of drought effected areas. There are desert development projects as well, but they are localised in the western part of the country, including the districts of Leh and Kargil in the state of Jammu and Kashmir. The total drought prone areas in the country amounts to about 260 million ha (Appendix II).

Floods

Another major environmental problem in the country is flood. The frequency and intensity of severe floods are increasing every year. In some parts of Ganga plains, there are up to three peak floods in a monsoon season. However, in some areas, medium or normal floods are welcomed as they are linked with the local agricultural practices. Occurrence of floods and droughts in the same place and time is becoming frequent. In Bihar plain the floods are due to excessive rainfall in the upper reaches, while climatic droughts conditions prevail over the same area. Though these problems continue to exist, the

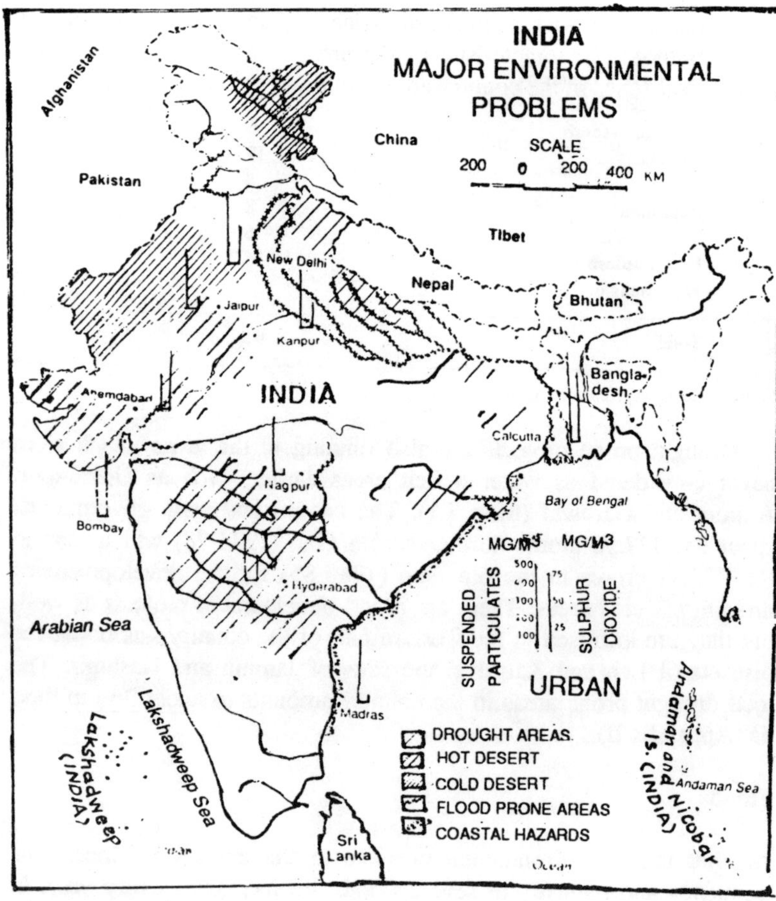

MAP 7.1: Major Environmental Problems

highest density of population is found in the flood prone areas. People have somehow adjusted with such environmental phenomenon. The total flood prone area in the country is about 40 million ha.

Commonly, floods occur either due to highly variable nature of monsoon, or due to cyclones and rain storms. But there can be local conditions as well, such as in coastal zone, or due to topographical and geomorphological characteristic of the area, or deforestation etc. The major flood zones in the country are as follows (Table 7.2):

TABLE 7.2: Distribution of area and catchments of major river valley projects and flood prone rivers.

Name	States	Total Area (in lakh ha)
A. *River Valley Projects*		
1. Bhakra (Sutlej)	H.P.	18.20
2. Chambal	M.P./Rajasthan	26.00
3. Damodar-Barakar/DVC	Bihar/W.B.	18.19
4. Hirakud (Mahanadi)	M.P./Orissa	83.20
5. Mahi (Kadana)	M.P./Rajasthan/ Gujarat	20.80
6. Matatila (Betwa)	U.P./M.P.	21.06
7. Nagarjunsagar (Krishna)	A.P./Karnataka/ Maharashtra	216.02
8. Nizamsagar (Manjira)	A.P./Karnataka/ Maharashtra	21.69
9. Pochampad (Godavari)	A.P./Karnataka/ Maharashtra	91.07
10. Rengali Mandira (Brahmani)	Orissa/M.P./ Bihar	25.25
11. Tunga Bhadra	Karnataka	28.28
12. Ukai (Tapti)	Gujarat/M.P./ Maharashtra	62.40
Total River Valley Projects catchments including 160 other small projects		693.23
B. *Flood Prone Catchments*		
1. Ajoy	Bihar/W.B.	610.00
2. Gomti	U.P.	30.14
3. Punpun	Bihar	8.50
4. Rupnaryan	W.B.	10.32
5. Son	M.P../U.P./Bihar	70.00
6. Sahibi	Rajasthan/Haryana/ Delhi	9.25
7. Upper Yamuna and Giri Bata	H.P./U.P./H.P.	14.15
8. Upper Ganga	U.P.	21.50
Total Flood prone Catchments		169.96
Total of A and B		863.19

Source: Indian Agriculture in Brief, Min. of Agriculture, New Delhi, 1986.

(a) Ganga plain and its tributaries, such as Yamuna, Ghaghra or Tista.
(b) Brahmaputra valley and its tributaries.
(c) Eastwards flowing deltaic rivers, such as Mahanadi, Godavari, Krishna and Cauvery.
(d) Westwards flowing rivers, such as Narmada and Tapti.

Measures have been taken to check the flood hazard. Multi-purpose river valley projects for flood control, irrigation and power generation were initiated after Independence, but the menance of floods is always increasing. Dams have created a new type of environmental risks in the lower reaches.

Associated Problems

Associated with droughts and floods are other problems which may lead to soil erosion and land degradation. These environmental problems are significant because they effect the agricultural production in this overpopulated country. The area degraded by different ways cover about 25 million ha which consists of waterlogged area (6 m ha), alkali soils (2.50 m ha), saline soil including coastal sandy areas (5.50 m ha), ravines and gullies (3.97 m ha), areas subjected to shifting cultivation (4.36 m ha), and riverine and torrents (2.73 m ha). Another major environmental problem is water and wind erosion. Out of 150 million ha area subjected to this type of erosion, 38.74 million ha (25.83%) is due to wind erosion and aridity. The latter figure also includes 7 million ha area covered by sand dunes. Average area annually subjected to damages through shifting cultivation is about one million ha. Further, there is about 3.98 million ha area which can be classified as gullies or ravines. The state of Uttar Pradesh has the highest 1.23 million ha (31%). Soil erosion and degradation is also due to shifting cultivation. Though no confirmed figures are available, such cultivation is mostly practised in the tribal areas of northeastern India and in Orissa. National Wasteland Development Board of the Ministry of Environment and Forestry has been entrusted in identifying different types of wastelands in the country so that a strategy to reclaim such land could be made.

The combined impact of the natural calamities multiplies the human sufferings. With mismanagement of such phenomena, the natural and human factors had to face catastrophe. Proper warning system or timely relief from the government could possibly reduce the impact of such events. The case of the small state of Haryana in northern India and Orissa in eastern India will reveal the pattern of occurrence of natural calamities during the past few years (Table 7.3).

TABLE 7.3: Natural calamities in Haryana (1979-88) and Orissa (1972-81)

Natural calamities	Haryana years	Orissa years
Drought	1979, 1980, 1981 1982, 1985, 1986 1987	1972, 1973, 1974 1976, 1977, 1978 1979, 1980, 1981
Floods	1980, 1983, 1985 1986, 1988	1972, 1973, 1974 1975, 1976, 1977 1978, 1979, 1980 1981.
Hailstorm	1982, 1987, 1988	
Cold wave	1984	
Cyclone		1972, 1973, 1974, 1975, 1977, 1979 1980, 1981
Thunder/Lightening		1972, 1974, 1975, 1977, 1978, 1980, 1981
Fire		1973, 1974, 1975, 1976 1977, 1978, 1980, 1981.
Storm		1974, 1976, 1978, 1979

Source: Revenue Department, Haryana and Statistical Abstract of Orissa.

In other states almost since 1977, droughts occur in 7 years out of 10 years. There have been spells of consecutive years of droughts. In Orissa, floods, droughts, cyclone and fire occur almost every year. There are other calamities as well. Such events have done extensive damage of crops. On the other hard, the frequency and severity of damages are increasing every year. Agriculture, settlements and population are continuously being exposed to new types of risks arising from the natural phenomenon.

Urban Environmental Problems

Recent concern for environment is not only due to natural phenomenon, but the urban, technological and industrial fall outs which have made people more aware of the environmental issues and their quality. Population beyond the multiple times of the sustaining capacity have made the urban centres polluted, overcrowded and created pressure on social and physical amenities. Concentration of industries in and around centres have exposed the urban population to all sorts of risks. These risks can remain unnoticed due to its slow introduction such as increasing quantity of pollutants in river Ganga, or can be sudden due to some industrial disaster. The case of Bhopal gas leakage in December 1984 is still fresh in our minds which was responsible for death of about 3,000 persons in the high density area around the Union Carbide pesticide plant.

The urban based environmental problems are as follows:

(a) Slums and housing conditions.
(b) Air pollution.
(c) Water pollution.
(d) Noise pollution.
(e) Solid waste disposal.

Slums, as a consequences of the prevailing socio-economic conditions, have now become an essential part of not only million cities of India, but also in medium size cities as well. According to one estimate about 20 per cent of the country urban population reside in the slums. This ratio is high in cities like Kanpur (40%), Lucknow (38.83%), Bombay (38.30%), Calcutta (35.35%) and Delhi (30.19%). Slum is not wholly an urban problem. Economically weaker section of rural population, such as landless labour, scheduled caste and tribe population and other social and economic backward communities tend to move to far away large urban centres like Calcutta or Bombay, or to nearby medium or small urban centres due to compulsion of various kinds in rural areas. Any attempts for betterment of urban environment makes urban centres more attractive thereby promoting rural-urban migration.

Air, water and noise pollutions have their origin in the urban-industrial-nuclear complexes. In some cases pollutants also effect the rural areas, particularly water pollutants. The excessive concentration of pollutants create conditions which are beyond tolerance limits and thereby become injurious to human and animal population, their safety, health and properties. In case of air pollution, the presence of sulphurdioxide (SO_2), Carbon monooxide (CO), fluorides, hydrocarbons, oxides of nitrogen and aldehydes, smoke and dust in large quantity makes sustenance of life dangerous in urban areas (Map 7.1). The source of these pollutants are vehicles, industries and wastes. In an air pollution survey over major cities of India, the highest quantity of suspended particulars (578 μgm^3) and the highest mean value of SO_2 (85 μgm^3) were found in Calcutta, followed by Bombay (275 & 83), Delhi (481 & 39), Kanpur (344 & 25), Nagpur (386 & 12), Jaipur (379 & 17), and Madras (106 & 16).

Water pollution, in a way, has more serious consequences because it can easily be transported and can enter in the food chain which may lead to genetic damages. The causes of such pollution may not be water based. Their origin may be owing to (a) atmospheric dissolved gases, (b) decomposition of animal and vegetable wastes, (c) industrial wastes, (d) sewage discharge, (e) weathering of rocks and soils, and (d) chemical, pesticide, radioactive and petroleum wastes. Presence of these wastes deteriorates the water quality because they stimulate the growth of micro organisms which often increase the biological oxygen demand (BOD) of the water. In 1986 a major project taken up by the Government of India in this regard was the cleaning of river Ganga under Ganga Action Plan. Since historical times there has been a concentration of urban centres along this river. Out of 47 urban centres along the river Ganga, there are 13 class I towns. In course of time, these towns have become sizeable by adding various types of industries and became multi-functional. The industrial and domestic wastes of these centres are flushed into this holy river. The disposal of dead bodies is an additional source of water pollution of the river. The cities like Allahabad, Kanpur, Varanasi, Patna and Calcutta are located along this river. Further Delhi, Agra and Lucknow are on the banks of the tributaries of the river Ganga. Action Plan for clearing other rivers are also underway.

Noise pollution and solid waster disposal are preliminary urban problems. Both these hazards increase with the density of population and the closeness to the centre of the city. Noise pollution can be from vehicles, loud speakers, music systems, crackers and the like. A survey carried out in some of the urban centres reveals that this pollution is high at road junctions, around railway stations, airport or bus stand. In general, there is a reasonable variation in the level of this pollution within a day. Acceptable levels of noise for various environmental set up have been worked out by the Indian Standards Institute (ISI). Similarly, the pattern of solid waste generated in cities is directly related to its population size. The urban garbage produced in India per day is about 0.35 kg per person. For Calcutta this average is highest, i.e. 0.57 kg per person. But, Bombay produces the maximum solid waste (3,000 tonnes per day), followed by Calcutta (2,500), Delhi (2,100), Madras (1,200) and Bangalore (1,000). Perception studies reveal that tolerance capacity of the population is much more than the scientifically accepted tolerance limits set for this purpose.

Environmental Protection Measures

Union and state governments are aware of the environmental issues. In each state, a department of environment has been established to monitor environment changes and to vet development projects. Remote sensing application in environmental studies has opened new vistas especially after successful launching of the Indian satellite IRS-IA. Laws are being passed in the parliament and state legislatures to protect the environment. These laws are to be forcefully enforced and observed. The Union Ministry of Environment and Forestry has recently prepared guidelines and questionnaires for environmental clearance of projects for six major sectors : (a) river valley, (b) thermal power, (c) industries, (d) shipping and harbour, (e) mining and (f) development of beaches. Biosphere reserves have already been identified. Ecologically fragile areas will be determined soon and their development programmes will be prepared.

The above guidelines provide a list ecologically or otherwise sensitive areas, like coastal areas, national parks and sanctuaries, estuaries, wetlands, archaeological monuments, flood plain and major human settlements, that may be avoided for setting up industries or development projects. It also enlists some aspects to be incorporated in the environmental management plans including treatment and disposal of liquid effluents, emissions of solid wastes and precautions for occupational safety. The guidelines for river valley and hydel power projects suggested that cost for environmental protection and mitigative measures should be included in the overall estimates to carry out cost benefit analysis. These projects should include measures like compensatory afforestation, restoration of land in construction areas, control of acquatic land, water and soil borne diseases and rehabilitation of project oustees.

Recent Considerations and Initiatives

It has widely been accepted that the problems in the field of environment in India is associated with poverty and under development. On the other hand, development activities were without environmental and ecological considerations. Now environmental management has become an essential component of development plans. In order to protect the environment, the government has taken some steps. It is likely that the proper implementation of certain acts will contributed to the improvement the quality of environment, rural or urban.

There are about 30 major acts by the central and state governments to protect or improve the environment. Some of the important acts are as follows:

(a) Water (Prevention and Control of Pollution) Act, 1974.
(b) Air (Prevention and Control of Pollution) Act, 1981.
(c) Factories Act.
(d) Insecticides Act.

The Environment (Protection) Act, 1986 provides a common basis of protect the environment and also to check the possible loopholes of the above acts. This act provide wide ranging powers to the Government to take necessary measures, confers power to individual to appeal in the court, makes obligatory to inform regarding discharge an apprehended discharge, and prescribes stringent penalties for the violation etc. Further the above mentioned Acts for Water (1974) and Air (1981) have been much more stringent.

On the other hand, institutions and bodies have been appointed to look after general or specific environmental problems. The Central Board for the Prevention and Control of Water Pollution is the apex body for prevention and control of water pollution. Till now 67 laboratories have been recognised as environmental laboratories in the country. Further, 173 monitoring stations all along the Indian Court have been identified for the purpose of water quality measurement. Particularly for River Ganga, Central Ganga Authority was set up in 1985 to oversee the implementation of the action plan drawn up for cleaning the polluted stretches of River Ganga.

It is now considered essential that the environmental issues are anticipated and included at the proposal formulation stage. All the development projects now include these aspects. In addition, attention has duly been given on eco-regeneration and development, biosphere reserves, wet lands, mangroves, natural resource management, waste land development, forest fire control, wild life, tiger project and environmental research, education and awareness programmes in collaboration with international scientific bodies, such as United Nations Environmental Programme, or International Union for Conservation of Nature and Natural Resources.

The environment related activities are governed and monitored by the Ministry of Environment and Forests and its especialised agencies. Other ministries and their departments are becoming environmental conscious as well. Government is supported by universities, voluntary agencies and activists. However, there is a lot of scope for development. The issues are enormous, and the challenge is linked with the survival of the mankind.

REFERENCES

Chowdhury, A.M.M. Dandekar and P.S. Pant (1989), Variability in drought incidence over India — A statistical approach, *Mausam*, Vol. 40, No. 2, pp. 207-14.

Kayastha, S.L. (1983), Floods in India — A study of their occurrences, causes, forecasting and control, *National Geographical Journal of India*, Vol. 29, Nos. 3 and 4, pp. 121-41.

Kayastha, S.L. (1986), Environmental politics in India, in Chris C. Park (ed.) *Environmental Policies: An Institutional Review*, Croom Helm, London, pp. 223-59.

Kayastha, S.L. and P. Nag (1989), The Bhopal Tragedy, in Clarke, J.I., P. Curson, S.L. Kayastha and P. Nag (eds.), *Population and Disaster*, Basil Blackwell, London.

Kumra, V.K. (1982), *Kanpur City: A Study in Environmental Pollution*, Tara Book Agency, Varanasi.

Nag, P. (1983), Perception of the environment: Issues and challenges, *National Geographical Journal of India*, Vol XXIX (Pt 3 and 4), pp. 217-31.

Reddy, N.B.K. (ed) (1979), *Drought Prone Areas of India*, Rayalaseema Geographical Society, S.V. University, Tirupati.

Singh, H.H., P. Nag, V.K. Kumra et al. (eds.) (1986), *Geography and Environment: Issues and challenges*. Concept Publishing Company, New Delhi.

Singh, O, V.K. Kumra and J. Singh (1988), *India's Urban Environmental*, Tara Book, Agency, Varanasi.

8

FORESTS, NATURAL VEGETATION AND WILD LIFE

The distribution of natural vegetation or forest is closely related to the distribution of rainfall. Forests are woody plants and as such a good indicator of physical, edaphic and climatic conditions of the land.

In India 7,43,534 km² area is under forests. Practically the whole of the forests area (91%) are government owned and controlled by the state forest departments. Most of the forests contain broad-leaved deciduous tree (85%) of which about one-sixth are *sal* forests growing in northern and central India, one-tenth teak forests occurring mostly in southern India. Teak and *sal* are the valuable timber trees in India. Coniferous forests occur mainly in the Himalaya between 2,150 and 3,850 m. The predominating trees are pines, junipers and *deodars*.

Area under Forest Administration

In the area under forest department is only 22.7 per cent of the total areas which less than 33 per cent as per the recommended of the National Forest Policy Resolution of 1952 (Map 8.1). There is very much variation in the proportion of forested land in different parts of the country (Table 8.1). The sparsely populated Andaman and Nicobar Islands are heavily forested, having more than 90 per cent of its land under forests. The inaccessible mountainous land of Arunachal Pradesh is also heavily forested. Here more than 60 per cent of its land are under forest cover. But in Jammu and Kashmir, a state in western Himalaya, the percentage of forest land is less than 10 per cent. In this state a large area lies above the tree-line and is too rugged at lower attitudes for the accumulation of soils. The densely populated plains of Punjab and Haryana, which are noted for green revolution, have cleared their forests to give room to arable land.

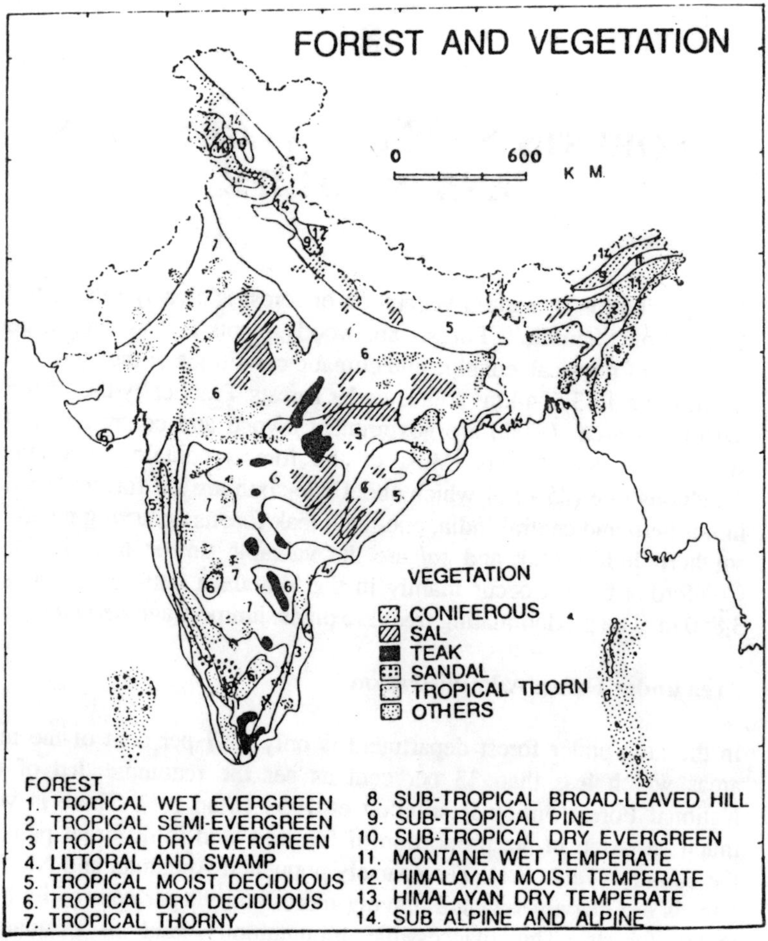

MAP 8.1: Forest and Vegetation

TABLE 8.1: **Forest types by states, 1981**

(*in percentage*)

States/UTs	Total Area	Coniferous	Non Coniferous	Merchantable
INDIA	22.62	5.68	85.0	76.28
Andhra Pradesh	23.47	—	100.0	78.54
Arunachal Pradesh	61.55	8.15	91.85	12.61
Assam	36.47	0.16	99.84	87.39
Bihar	16.84	—	100.00	100.00
Gujarat	9.33	—	100.00	94.86
Goa, Daman, Diu	N.A.	N.A.	N.A.	N.A.
Haryana	3.24	25.59	74.41	79.29
Himachal Pradesh	38.92	49.28	50.72	57.78
Jammu and Kashmir	9.47	91.74	8.26	40.72
Karnataka	19.34	—	100.00	75.72
Kerala	28.99	—	100.00	95.69
Madhya Pradesh	36.62	—	100.00	99.67
Maharashtra	21.73	—	100.00	75.46
Manipur	26.97	15.06	84.97	60.64
Meghalaya	37.19	9.89	90.11	53.20
Mizoram	N.A.	N.A.	N.A.	N.A.
Nagaland	17.35	9.00	91.00	81.99
Orissa	43.62	—	100.00	70.03
Punjab	4.20	6.05	93.95	47.22
Rajasthan	9.96	—	100.00	83.37
Sikkim	N.A.	N.A.	N.A.	N.A.
Tamil Nadu	17.34	0.02	99.98	94.78
Tripura	57.20	—	100.00	80.74
Uttar Pradesh	17.42	10.77	68.84	60.61
West Bengal	13.33	0.77	99.23	94.99
Andaman & Nicobar Islands	90.48	—	100.00	100.00

Source : S.P. Chatterjee, *Economic Geography of Asia*.

In Indian forests, important fact is the preponderance of non-coniferous timber trees over coniferous trees. The commercial viability of forests depend on their ready accessibility and usefulness of their produce, and the turn out of forest products in terms of timber, pulp, match wood, round wood, firewood and charcoal wood. Besides supplying timbers and other forest products, India's forests also help in minimising soil erosion and to provide shelter belts and wind belts. Their conservation and utilisation on a planned basis is therefore a major economic necessity.

Classification of Forests

A more recent study of Indian forests have been classified India into 14 major types of forests (Map 8.1) of which six forests are tropical, three subtropical, three temperate, one subalpine and alpine and one is littoral and swamp.

Tropical Wet Evergreen and Moist Semi-Evergreen Forest

These forests occur in areas of heavy rainfall exceeding 2,500 mm. A strip along the Sahyadris up to a height of 1,370 m and large areas in Assam, Meghalaya, Tripura, Manipur and Andaman Nicobar Islands up to a height of 1,100 m are covered with these forests. The wet evergreen forest is bordered by the semi-evergreen on the drier side which in turn merges with the tropical moist deciduous forests. The vegetation in these forests is very dense and trees often reach 45 m in height. The ground is covered by a tangled mass of cones, palms, bamboos and ferns. The species are numerous and valuable which include, rosewood, *paan*, *champa*, *gurjan* etc.

Tropical Moist and Dry Deciduous Forests

These forests are called the monsoon, forests which form the natural cover over nearly whole of India, between the Himalayas, the Thar and the Western Ghats. These areas have moderate rainfall varying from 1,000-2,000 mm. The moist deciduous forests form a long strip along the Sahyadris occupying most of Orissa. eastern Madhya Pradesh, Chota Nagpur plateau and strip along the Siwaliks including the Bhabar and the Terai where the rainfall is between 1,500-2,000 mm. Economically, these areas have the most important forests of India. Sal is the most significant variety. Teak is the characteristic of the monsoon forest.

Tidal Forest

On the tidal washed coasts, dense mangrove forests flourish. The seaward fringes, channels and islands of the deltas of the Ganga, Mahanadi, Godavari and Krishna rivers are belts of dense tidal forests which grow in brackish as well as fresh water. The great Sunderban delta is covered by the Sundari trees.

Tropical Dry Evergreen Forests

This forests is confined only to Tamil Nadu coast. It has a closed but low canopy of grasses and shrub.

Thorn Forests

In areas with under 750 mm of rainfall the natural vegetation consists of open stunted forests breaking down into xerophytic bush and in the north western grading into deserts. Trees are few and are widely scattered. Acacia are most important and associated with thorny shrubs. The vegetation is predominant in the Indus plains, Rajasthan and also in many places of interior peninsula.

Montane Subtropical and Temperate Forests of Southern India

These includes the wet hill forests and wet temperate forests of the southern hills of Peninsular India. Nilgiris and the Palni hills in the south have a wet hill forests between 1,000-1,500 m in altitude. It is mainly a stunted type of rainforest which occur below it. Above 1,500m on the slopes of the Nilgiris, the Annamalais and palnis occur the wet temperate forest. The forests is dense but low with much undergrowth and many epiphytes, mosses and ferns are common.

Wet Hill Forest

A fairly dense and high wet hill forest occur at 1,000-2000 m in height in the eastern Himalaya. Evergreen, oak, chestnut, ash, beech at a lower level are common.

Subtropical Forests

These forests occurs at a height of 1,000-2,000 m in the western Himalayas. Chif is the most important tree while evergreen and oaks occur in the wetter areas.

Subtropical Dry Evergreen Forests

These forests occur in the restricted areas on the Himalayan foothills in

Kashmir in the areas where the rainfall varies from 500-1,000 mm. The forests contain wild olivers with varieties of scurb.

Moist Temperate Forests

These forests occur in the entire Himalayan region between altitudes of 1,500-3,300 m. In the areas rainfall varies from 1,000-2,500 mm. Evergreen, oaks, laurel and chestnuts and dwarf bamboos are commonly found here.

Dry Temperate Forests

These forests are found in the main Himalayan ranges where rainfall is below 1,000 mm. These forests are open and contain xerophytic trees with *deodar*, juniper, birch and older as dominating trees.

Alpine Forests

The Himalayas from 3,000-3,600 m in height are covered by a dense scruby forest of silver fire juniper, pine, birch and rhodedendron.

It is very shocking to note that there is less awareness among Indian mass regarding the importance of forest and vegetation. Mass deforestation due to some social, economic and cultural reasons is creating environmental imbalances. It is necessary to create an awareness among the people regarding the necessity of forest. The deforestation is dangerous for wild life preservation and protection as well.

Wild Life

The wild life of South Asia is extremely varied, comprising of all the major groups of the animal kingdom. Nearly 76 thousand species of animals live here which represent 82 per cent of the known living species of the world.

India still abounds in wild life, though their numbers have considerably decreased in recent times as a result of excessive hunting and opening up of forest land. Of the wild animals, elephants, bisons, bears, panthers and tigers still roam in hilly and forested lands of the country. Lions, rhinoceroses, hunting leopards and the *kasturi* stages have decreased in number so much that their hunting and export from

India have been prohibited by the Government of India. Wild life sanctuaries and national parks have been set up in different parts of the country for the preservation of wild animals. Gir forest in Gujarat, is the only sanctuary in the world for the preservation of lion. Jaldapara sanctuary in West Bengal, Kaziranga sanctuary in Assam and Corbett National Park in the Siwalik hills of Uttar Pradesh are some of the well-known places set up by the Government of India for the preservation of wild life. Wild sheep and goats, the markhov, the ibex, the scrow and the takin are found in the Himalayan ranges. Birds of several species especially pheasants, geese, ducks, mynahs, parakeets, pigeons, cranes, hornbils and sun birds are still found in large numbers.

Many wild life species are on the verge of extinction, such as the mountain quail and some species of reptiles. The Wild Life (Protection) Act, 1972 adopted by most of the Indian states, governs the wild life conservation and protection of endangered species, both inside and outside forests.

REFERENCES

Ali, S. (1961), *The Book of Indian Birds*, Bombay.

Kayastha, S.L. (1970), Conservation of natural resources in Himalaya — a vital need, *National Geographical Journal of India*, Vol. 16 (3 & 4), pp. 208-20.

Ministry of Agriculture and Irrigation (1976), *Report of the National Commission on Agriculture*, Part IX, Forestry, Government of India, New Delhi.

Puri, G.S. (1960), *Indian Forest Ecology*, Oxford Books & Stationary, Delhi, (2 Vols.).

Singh, A. K., V.K. Kumra and J. Singh (1987), *Forest Resources, Economy and Environment*, Concept Publishing Company, New Delhi.

9

RESOURCES

India is a fortunate to have abundant natural resources which are essential to develop the country's economy. The transformation of traditional society into modern one is possible with the help of energy, and mineral resources. India needs proper technology to use these resources to make the development smooth and profitable.

Electricity

Electric power is a key resource in the development process. Coal and water power contribute primarily to the generation of energy in India, but these resources have their limitations. Conventional thermal power stations require large quantities of coal for their operations. Since the larger coal fields are situated in three states — Bihar, West Bengal and Madhya Pradesh, the long haulage distances hamper the economic of power generation. The hydro power resources in India are evenly distributed (Table 9.1). The greater advantage of its use as a source of power is that it provides cheap power since the running costs are the lowest and the maintenance of power station is also easier. Nuclear power provides much smaller quantities of power because of the paucity of nuclear minerals in India. Moreover, the nuclear power stations require heavy investment and the maintenance, and operational problems require co-ordination of technicians from different disciplines ranging from nuclear physics to instrumentation. The operational problems of nuclear power stations are far more complex than those posed by conventional hydro and thermal power stations. The main advantage of nuclear power station is that it does not require large quantities of fuels and is not affected by transport bottlenecks.

The total installed capacity of electric energy in India in 1981-82 was 12,210 crore kWh of which 6,874 crore kWh was generated by thermal power stations, 4,956.5 crore kWh by hydro power stations, and 302.1 crore kWh by nuclear power stations.

TABLE 9.1: Regionwise generation of electricity

Region	1984-85	1987-88
A. *Northern*	41,507	46,149
Thermal	22,027	25,178
Hydro	18,402	19,679
Nuclear	1,078	1,292
B. *Western*	51,700	57,389
Thermal	42,980	49,486
Hydro	6,790	5,941
Nuclear	1,930	1,962
C. *Southern*	42,132	43,486
Thermal	16,323	20,400
Hydro	24,739	31,355
Nuclear	1,070	1,731
D. *Eastern*	19,635	21,175
Thermal	16,592	18,215
Hydro	3,043	2,960
E. *North-Eastern*	1,659	1,838
Thermal	848	840
Hydro	811	998

Source: Central Electricity Authority.

Power Generation

In India there is chronic deficiency in electric energy resulting in load shedding and dislocation of industrial activities. The regional break up of energy generated in 1981-82 shows that the Southern Regional Electricity Board comprising states like Andhra Pradesh, Karnataka, Tamil Nadu, Kerala and Pondicherry depended more on hydro power (73%) because of the dearth of mineral fuels, especially coal in this region. In contrast, the Eastern Regional Electricity Board comprising three states, i.e. Bihar, West Bengal and Orissa including the jurisdiction of the Damodar Valley Corporation generated more (70%) of coal based thermal power, because of the abundance of coal fuel in this region. The Northern Regional Electricity Board comprising the six states of Jammu and Kashmir, Himachal Pradesh, Punjab, Haryana, Uttar Pradesh, Rajasthan, and two union territories of Chandigarh and Delhi gave almost equal stress on hydro (57%) and thermal (43%)

power in the generation. The Western Regional Electricity Board comprising three states, i.e. Madhya Pradesh, Gujarat and Maharashtra utilising more thermal power (74%) than of hydro power (26%). The North Eastern region comprising seven states of Assam, Meghalaya, Nagaland, Manipur, Tripura, Mizoram, and Arunachal Pradesh. They generated least amount of electricity (120.1 crore kWh) and this region use gas and oil for the generation of power along with thermal and hydro power sources.

Power Corporations

The Government of India has set up National Thermal Power Corporation (NTPC) and National Hydroelectric Power Corporation (NHPC) in the later part of 1970s to tap hydal power and to distribute both, thermal and hydal power, throughout the country with the help of the national grid system (Table 9.2 A&B). These corporations work along with the state electricity boards and other electricity generating organisations. Though hydro electric power has been generally the most economic source of power but the progress of the production of hydro power has not been as fast it should have been. The initial capital cost of hydro power generation is very high because it requires construction of dams in hilly terrains and then transmission of power in different parts of the country. It is also true that the construction dam in mountainous region creates some environmental imbalances hence proper attention should be given on the selection of the site of the project. Otherwise, these projects will become useless in future mainly due to the siltation in the dams.

Atomic Energy

India embarked on the atomic energy programme on the December 2, 1942 and has been steadily solving many of the problems with which the Indian Atomic Energy Commission confronted from time to time. As the natural uranium resources in India are limited, the Atomic Energy Commission's first task was to minimise the total quantum of power generated by heavy water reactors and maximise the contribution of the fast breeders. The installed capacity of nuclear power plants in India is 860 crore kWh and the energy generated is about 300 crore kWh.

TABLE 9.2: Thermal and hydro-electric ower projects

A. National Thermal Power Corporation Project

Projects	Total (MW) Capacity	Present (MW) generation
1. Singravli	2,000	1,000
2. Korba	2,100	600
3. Ramagundam	2,100	600
4. Farakka	1,600	600
5. Vindhyachal	1,260	600
6. Rihand	1,000	500
* 7. Kahalgaon	840	
* 8. NCTPP/Dadri	840	
* 9. Badarpur	210	
*10. Talcher	1,000	

B. National Hydro-electric Power Corporation Projects

Projects	Capacity (MW)
1. Baira Sivl (HP)	750
2. Lak Tak (Manipur)	409
3. Salal (J & K)	533
* 4. Chemera (H.P.)	540
* 5. Dul Hasti (J & K)	390
* 6. Tanakpur (UP)	120
** 7. New Uri (J & K)	480
** 8. Chamera II (H.P.)	300
** 9. Rangit (Sikkim)	60
**10. Baglibar (J & K)	450
**11. Swalkot (J & K)	600
**12. Dhavliganga (U.P.)	210
**13. Salal II (J & K)	345
**14. Dhalswari (Mizoram)	120

* On going
** Proposed

Source: NTPC Annual Report 1988-99.
NHPC Annual Report 1988-89.

Mining Sector

India is fairly rich in many minerals, most of which are exploited in large quantities. The country has 4,181 (1983) mines distributed all

over India. At least four minerals are in great demand in the world markets: iron ore. mica, titanium ore and thorium ore. India has also exportable surplus in many other minerals, such as bauxite, manganese ore, magnesite, silimanite and kyanite. The country is self-sufficient in coal, building stones and cement raw materials (Appendix III). Gold, silver, precious and semi-precious stones, such as diamond, emerald and sapphire are also available in India (Table 9.3).

TABLE 9.3: Mineral production

Mineral	1971	1981	1987
			('000 tonnes)
Bauxite	1,517.0	1,785.0	2,685.0
Chromite	275.0	321.0	642.0
Copper ore	666.0	2,005.0	4,985.0
Gypsum	1,089.0	866.0	1,861.0
Lead	4.3	16.8	48.0
Manganese Ore	1,841.0	1,695.0	1,303.0
Mica	15.0	7.9	4.0
Zinc	15.9	46.5	102.7
			(million tonnes)
Coal	71.8	109.1	175.6
Iron Ore	34.3	41.9	52.0
Limestone	25.1	29.2	57.4
Petroleum (crude)	7.2	9.4	30.8
Gold (Kg)	3656	2452	1863

Source: Statistical Outline of India 1988-89.

In terms of value of output, coal is the foremost mineral. Most of the coal bearing mines occur in the Damodar Valley region of Bihar and West Bengal. Next comes in importance the Singrauli coalfield in Madhya Pradesh. Coal deposits occur in the Godavari Valley which is the main source of supply to the southern states.

India's production of petroleum (crude) has been increasing steadily. The production from the older petroleum fields in Assam being supplemented by the production from the newer oil fields in Gujarat. The half of the iron ore in India was mined in Singhbhum and Mayurbhanj districts of Bihar and Orissa and the other half was produced in Madhya Pradesh, Maharashtra, Goa and Karnataka. India is the largest producer of manganese ore in Asia. The principal manganese ore mining areas are located in Madhya Pradesh, Bihar, Orissa, Andhra Pradesh, Goa, Karnataka. Smaller copper ore deposits are found scattered in peninsular India and in the Himalayan region, but workable larger deposits occur in Bihar, Rajasthan and in Andhra Pradesh. Bauxite is the principal ore of aluminium and is also used in the manufacture of alloys and chemicals, refractories and abrasive. The main producing centres are located in Maharashtra and Madhya Pradesh. The gold in India comes from the Kolar Gold Fields and Huti mines in Karnataka. Lead ore deposits occur in Andhra Pradesh and Orissa. The present production of tungsten ore, a strategic mineral comes only from West Bengal and Rajasthan.

India's atomic minerals consists mainly of pitchblende, a source of uranium; and monazite, a source of thorium. Bihar and Rajasthan produce most of the uranium and the beach sands on the east and west coasts produce most of thorium.

India also produces a large number of non-metallic minerals. Some of the major non-metallic ores are mica, asbestos, appetite and phosphorite, magnesite, silimanite, kyanite, gypsum and pyrite, tarytes, limestones, dolomite, and rock salt. India is the largest producer of mica in the world. Moreover, Indian micas are of good quality and are in great demand abroad. Bihar has the richest mica belt. The other two mica belts are in Andhra Pradesh and Rajasthan. More than eighty per cent of the country's production of asbestos comes from Rajasthan.

In India the phosphatic deposits occur principally as appetite and phosphorite which served as raw material for manufacturing chemical fertilisers. Phosphorite deposits occur in Uttar Pradesh and Bihar. Rock phosphate has an advantage of being directly used as fertiliser in the fields. Appetite deposits of Andhra Pradesh and Bihar and rock phosphate of Rajasthan, Uttar Pradesh, Madhya Pradesh are commercially exploited for this purpose. Andhra Pradesh is endowed with vast resources of lime stone suitable for the manufacture of cement. Dolomite is increasingly used in steel industry. Madhya Pradesh and Orissa are the principal producing states.

Magnetite, sillimanite and kyanite are refractory minerals used in building furnaces for the metallurgical industry. Meghalaya, Maharashtra have most sillimanite deposits. Bihar and Maharashtra have almost monopoly of kyanite. More than 80 per cent of India magnetite comes from Tamil Nadu and Karnataka.

India's requirements for sulphur, an essential ingredient for the fertiliser and chemical industry, are met by deposits of gypsum and pyrites, entire production of former coming from Bihar and four-fifths of latter coming from Rajasthan. Andhra Pradesh has most of India's barytes limestones and dol mites serve the purpose of both rocks and minerals. They are used not only as building stones but also as raw materials for lime and cement manufacturing and also as a flux in iron and steel industry. Madhya Pradesh and Tamil Nadu lead in their production.

India's precious and semi-precious stones consists of diamonds emeralds, rubbies and sapphires. The whole of diamond comes from Madhya Pradesh emeralds and sapphire mines are located in Rajasthan and Kashmir.

Minerals form one of the modern world's important resources in the support and fulfilment of human life. The extensive use of minerals in some cases has caused depletion of the the stock of minerals. The diminishing reserves of minerals like lead, tin, zinc, copper, nickel and eventually of petroleum and coal have attracted attention of geologists and industrial economists. Hence a planned programme of mineral development in India is required. The national economy demands that India's resource production and its conversion into industrial manufactures within the country should bear proportion to the volume of her main natural resources. A long term strategy is required regarding the conservation, use and refuse of the limited resources such as minerals (Map 9.1.)

Water Resources

Different aspects of water resources have been described in the chapters on physiography and drainage, environment and irrigation. This resource has been mostly evaluated in relation with its utilisation in agriculture, which is the backbone of the country's economy. For the same reason, floods and droughts have received much attention as in these cases the available water is either in excess or in shortage for agriculture. But recently, with rapid urbanisation, domestic requirement of water is increasing at a very faster rate. Multiplication of permanent dwellings in urban areas are connected with tap water supply, demand of which exceeds the reservoir capacities meant for this purpose. Considering this trend, urban drought has been predicted by the turn of the century in India. The water shortage is already been experienced in cities like Madras, Delhi or Dehra Dun during the summer season.

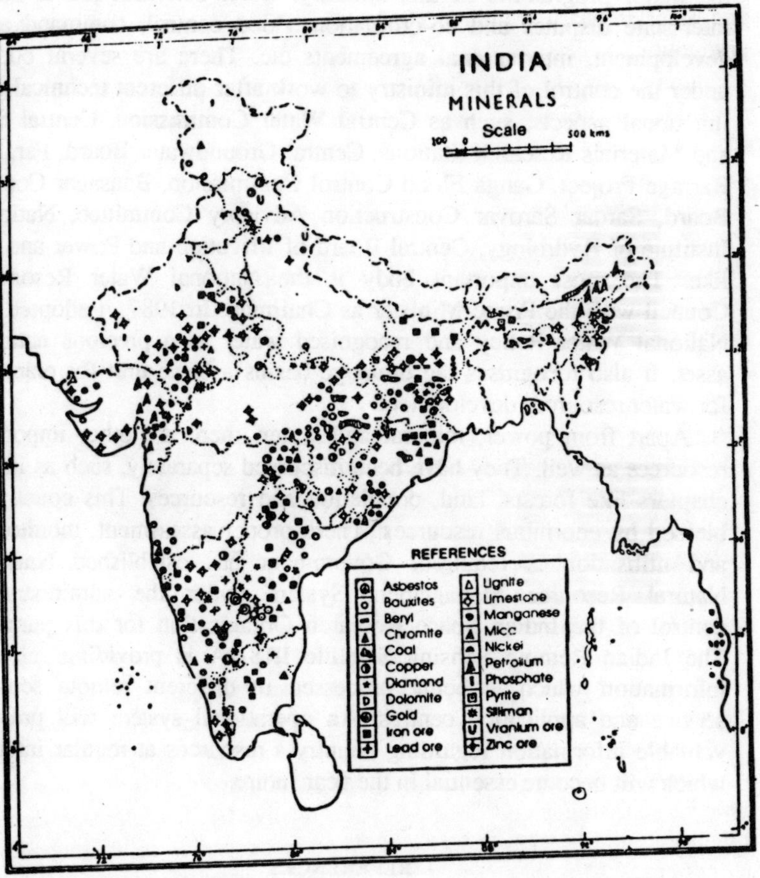

MAP 9.1: Minerals

Considering the importance of this resource, Ministry of Water Resources has been formed. It is responsible for broadly laying down the policies and programmes for the development and regulation of water resources, both surface and groundwater. Irrigation is of course the major programme of this ministry. Other activities are to solve inter-state disputes and co-ordination, flood control, command area development, international agreements etc. There are several bodies under the control of this ministry to work after different technical and functional aspects; such as Central Water Commission, Central Soil and Materials Research Stations, Central Groundwater Board, Farakka Barrage Project, Ganga Flood Control Commission, Bansagar Control Board, Sardar Sarovar Construction Advisory Committee, National Institute of Hydrology, Central Board of Irrigation and Power and the like. The most important body is the National Water Resources Council with the Prime Minister as Chairman. In 1987, it adopted the National Water Policy and recognised water as a precious national asset. It also recognises the drainage basins a basic unit for planning for water resources development.

Apart from power, mineral and water, there are other important resources as well. They have been discussed separately, such as in the chapters like forests, land, population and resources. This country is blessed by enormous resources. There proper assessment, monitoring and utilisation is required. Government has established National Natural Resource Management System under the administrative control of the Indian Space Research Organisation for this purpose. The Indian Remote Sensing Satellite IRS-1A is providing relevant information which is being processed in different remote sensing service and application centres. An operational system will provide valuable information regarding country's resources at regular interval which will become essential in the near future.

REFERENCES

Chaudhuri, M.R. (1970), *Power Resources of India*, Oxford & IBH Publication Co., Calcutta.

Kayastha, S.L. (1981), An appraisal of water resources of India and need for a national water policy, *Geo Journal*, Vol 6, pp. 563-72.

Kayastha, S.L. (1975), Resource appraisal in India, in Walter Imber H. Boesch (eds.), *India*, Kummerly and Frey, Berne, pp. 106-19.

Kayastha, S.L. (1981), Energy problems in India — Socio-economic aspects, in R.P. Chaturvedi et al. (eds), *Socio-Economic Aspects of Energy*, Catland, New York.

NATMO (1990), *Water Resources Development Atlas of India*, NATMO, Calcutta.

Rao, K.L. (1975), *India's Water Wealth*, Orient Longman, New Delhi.

Raza, Moonis (ed.) (1989), *Renewable Resources for Regional Development*, New Delhi.

Roy Burman, B.K. (1970), *Resources Region of India*, Office of the Registrar General, New Delhi.

Rehman, Hifzm (1985), Trends in energy consumption in India, *The Geography*, Vol. XXXII(1), pp. 36-51.

Singh, R.M. (ed.) (1987), *Coal Resources of India*, Department of Geology, BHU, Varanasi.

Sengupta, S. (1988), Problems and Potentials of non-conventional energy resources in India, *NATMO Monograph 8*, Calcutta.

10

LANDUSE

The variation in the topographical features in combination with the human and technological inputs has been responsible for a variety of uses of land in the country. In urban areas the landuse is more complex due to the multiple uses. 'The success of multiple landuse depends upon the compatibility of the different uses and upon the land and quality of resource management'. Same piece of land in rural areas can be used for several purposes, for example water bodies can be good for fishing and recreation. However, standard landuse categories have been identified throughout the country (Map 10.1). All India totals for each category of land is now available by adding different uses of land in villages and towns of the country. But, there is substantial area for which no returns exists. The percentage of such area is reducing from 13.5 per cent in 1950-51 to 7.5 in 1982-83 (Table 10.1).

Area under Forest Administration

The area under the forest administration has increased from 14.2 to 22.1 per cent since Independence, but the actual forest cover has reduced substantially, party due to development projects like dams. On the other hand, the ever-increasing population pressure has been responsible for the reduction of area not available for cultivation, either non-agricultural use or barren or uncultivable land. Similar trend has also been noticed for 'other uncultivated land' which include pasture, grazing lands, miscellaneous tree crops, groves and cultivable wastelands. Regarding fallow lands, the current fallow has increased but other fallow land has decreased. The net area sown and total cropped has increased indicating higher cropping intensity since Independence.

TABLE 10.1: India: Landuse changes (% of the reporting area)

Landuse Category	1950-51	1960-61	1970-71	1982-83
1. Area under forests	14.2	18.1	21.0	22.1
2. Area not available for cultivation	16.7	17.0	14.7	13.0
(i) Area under non-agricultural use	3.3	5.0	5.4	6.4
(ii) Barren & unculturable land	13.4	12.0	9.3	6.6
3. Other uncultivated land (excluding fallow land)	17.4	12.6	11.6	10.5
(i) Permanent Pasture & other grazing land	2.3	4.7	4.4	3.9
(ii) Land under misctree crops and groves not included in net area sown	7.0	1.5	1.4	1.2
(iii) Cultivable wasteland	8.1	6.4	5.8	5.4
4. Fallow lands	9.9	7.7	6.4	7.8
(i) Fallow lands other than current fallow	6.1	3.8	2.9	3.1
(ii) Current fallow	3.8	3.9	3.5	4.7
5. Net area sown	41.8	44.6	46.3	46.6
6. Total cropped area	46.4	51.2	54.6	56.8
7. Area sown more than once	4.6	6.6	8.2	10.1
8. Cropping intensity*	111.1	114.7	117.8	121.8
9. Total area reporting (000 ha)	284,315	298,458	303,758	304,110
10. Area for which no return exists	13.5	9.2	7.6	7.5

$$\text{* Cropping intensity} = \frac{\text{Total Cropped Area}}{\text{Net Area Sown}} \times 100$$

Source: Indian Agriculture in Brief.

A comparative picture of different states and the union territories in total shows that the latter have the largest percentage of land under forests and under the control of forest departments (80.79%) (Table 10.2). Among the states, Jammu and Kashmir has 61.23, Tripura has 55.15, Orissa has 42.73 and Sikkim has 36.35 per cent area under the control of forest departments. The well populated state of Kerala is an exception. Here 27.85 per cent of area is under forest control. On the other hand, there are states which has less than ten per cent area under forests which is much less than the desirable limits of 33 per cent. These states are Haryana, Punjab and Rajasthan. There are districts in other states where no forests exist, while in some states there is a lopsided distribution of forest. For example in Uttar Pradesh, only the Himalaya and Mirzapur areas are having substantial forest cover. Further, the areas under forest departments do not necessarily mean that it is covered by forests. In such areas, settlements grazing land, degraded forests or simply unused land may exist. Forests have been considered as an essential element for maintaining the ecological balance. Hence, lately a lot of attention has been paid to monitor this natural resource (Map 10.1)

Non-agricultural Uses

The area under non-agricultural is 6.44 per cent in India. It is 14.62 per cent in West Bengal and 13.69 in Tamil Nadu. Excepting Assam, the northeastern region has low percentage of land in this category. Though, barren or uncultivable land have similar percentage, but the variations are larger. Manipur has as high as 64.18 per cent in this category and along with forests over 90 per cent of the land in this state is under non-agricultural uses. However, the population density is only 64 persons to a square kilometre. Sikkim and Assam also have reasonable area under this category.

Due to the nature of economy, Himachal Pradesh and Sikkim have sizeable percentage of area as permanent pastures or grazing land. Sheep-rearing has been an important source of employment in these states. On the other hand, in plain areas, the land under this category is negligible. The exceptions are Rajasthan, Karnataka and Madhya Pradesh. Further, in the hilly areas and in the northeastern region, another category of landuse which may not be considered under agriculture *per se* but the income from the use of land can either be a supplementary or secondary source of rural income, i.e. miscellaneous tree crops or groves, has been found sizeable. In Nagaland, it is over 18 per cent. However, in western states like Gujarat and Rajasthan, the area categorised as culturable wasteland is over ten per cent. Meghalaya has over one-fifth area in this category. In the same state large areas are also fallow. The situation of current fallow is different and likely to change fast. However, in the year 1982-83, in Nagaland, Tamil Nadu and Bihar extensive areas were reported under current fallow.

TABLE 10.2: Landuse 1982-83

State/UTs	Reporting Area in 000 ha	Landuse as percentage of reporting area												
		Forests	Non-Agricultural use	Barren Uncultu-rable land	Permanent Other grazing land	Misc. tree Crops/ groves	Culturable wasteland	Fallow lands other than current fallow	Current fallows	Net area sown	Total cropped area	Net irri-gated area	Gross irri-gated area	Cropping inten-sity
1	2	3	4	5	6	7	8	9	10	11	12	13	14	15
Andhra Pradesh	27,440	22.49	8.08	8.37	3.28	0.99	3.28	5.35	7.95	40.21	46.53	12.85	16.47	115.70
Assam	7,852	25.28	11.61	19.63	2.34	3.18	1.36	1.09	1.16	34.35	45.29	7.28	7.28	131.90
Bihar	17,330	15.79	9.57	5.56	0.78	1.23	2.39	5.41	11.49	47.77	55.62	13.38	19.33	11.65
Gujarat	18,826	10.43	5.73	13.29	4.49	0.02	10.46	1.60	2.61	51.37	54.12	11.47	13.90	105.4
Haryana	9,394	5.37	7.46	2.00	0.61	—	1.09	—	3.89	81.84	120.76	53.61	80.97	147.6
Himachal Pradesh	3,113	25.99	4.91	4.63	34.9	1.32	7.68	0.45	1.38	18.34	30.77	2.99	5.10	167.8
Jammu Kashmir	4,506	61.23	6.75	5.68	2.73	2.08	3.22	0.18	2.06	16.07	22.24	7.06	8.99	138.4
Karnataka	19,050	15.9	5.94	4.35	6.61	1.74	2.60	2.57	5.91	54.36	58.54	7.80	9.41	107.7
Kerala	3,885	27.85	7.10	2.21	0.13	1.42	3.35	0.69	1.13	56.11	73.67	6.67	10.01	131.3
Madhya Pradesh	44,211	31.70	4.98	5.36	6.38	0.37	4.60	2.28	1.83	43.04	50.25	6.01	6.21	116.8

(Contd.)

Contd.)

1	2	3	4	5	6	7	8	9	10	11	12	13	14	15
Maharashtra	30,758	17.27	3.24	5.59	5.15	0.59	3.20	2.61	2.82	59.50	72.23	6.27	8.73	109.0
Manipur	2,211	27.23	1.18	64.18	—	1.09	—	—	—	6.33	8.50	2.94	3.39	134.3
Meghalaya	2,249	36.10	3.78	10.27	0.76	6.45	20.19	11.61	2.27	8.58	9.25	2.22	2.27	107.8
Nagaland	1,099	26.02	2.54	—	—	18.20	5,73	5.73	23.84	8.64	15.00	6.01	6.92	107.9
Orissa	15,540	42.73	4.07	1.71	3.60	2.72	1.60	1.22	2.91	39.45	53.58	7.82	12.91	135.8
Punjab	5,033	4.29	8.74	1.79	0.08	0.08	0.74	0.02	0.77	83.49	137.39	70.53	122.15	164.6
Rajasthan	34,235	6.28	4.41	8.45	5.36	0.15	18.09	5.61	5.90	45.74	53.73	9.40	11.94	117.5
Sikkim	729	36.35	9.60	28.67	13.31	0.27	0.14	0.96	0.27	10.43	14.13	1.65	1.65	135.5
Tamil Nadu	13,001	15.61	13.69	4.54	1.19	1.55	2.54	4.37	16.05	40.45	46.38	17.34	21.01	114.7
Tripura	1,048	55.15	—	11.45	—	9.35	0.19	0.19	0.19	23.47	36.35	2.76	3.63	154.9
Uttar Pradesh	29,748	17.21	7.85	3.76	1.01	1.91	3.86	2.54	3.95	57.91	83.06	33.23	40.76	143.4
West Bengal	8,846	13.38	14.62	1.31	0.05	1.83	4.23	0.69	0.93	62.91	79.18	20.73	20.73	125.9
Union Territories	9,006	80.79	1.23	0.31	0.13	0.77	2.55	3.65	2.45	5.25	6.59	1.37	1.64	125.6
All India	3,04,110	22.08	6.44	6.62	3.93	1.17	5.36	3.12	4.65	46.62	56.77	13.14	17.11	121.8

Source: Indian Agriculture in Brief.

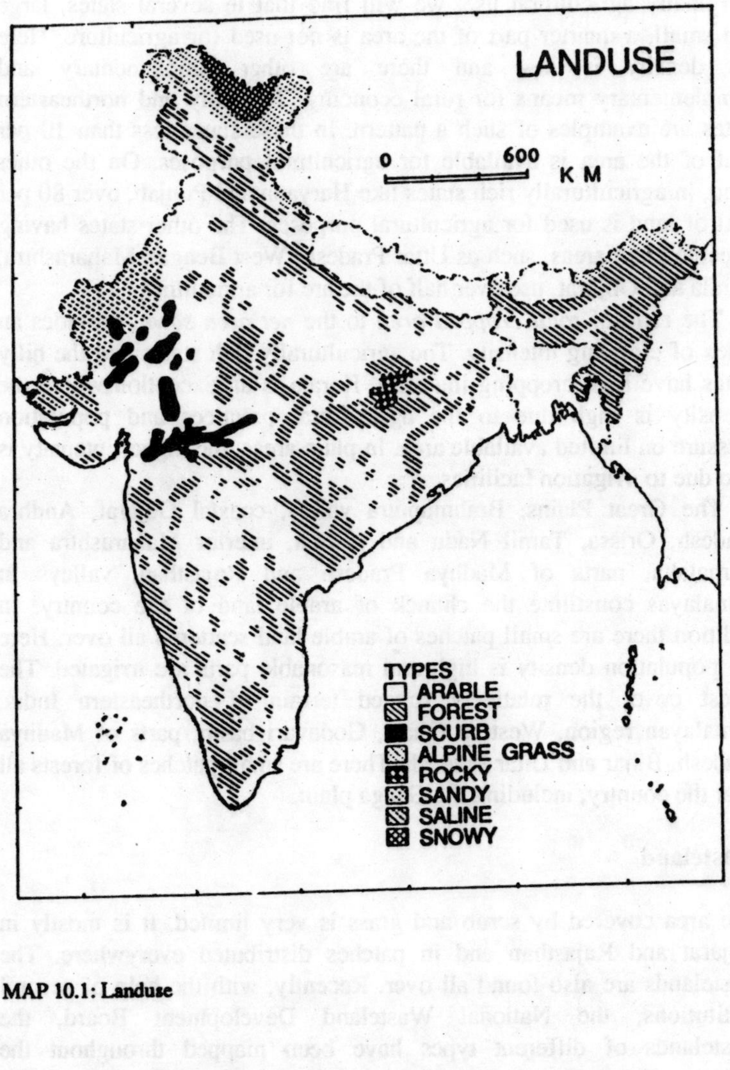

MAP 10.1: Landuse

Agricultural Uses and Cropping Intensity

Considering the above pattern of landuse of non-agricultural or marginally agricultural use, we will find that in several states, large and small, a quarter part of the area is not used for agriculture. Here the density is low and there are other supplementary and complementary means for rural economy. The hilly and northeastern states are examples of such a pattern. In these states less than 10 per cent of the area is available for agricultural purposes. On the other hand, in agriculturally rich states like Haryana and Punjab, over 80 per cent of land is used for agricultural purposes. The other states having sizeable plain areas, such as Uttar Pradesh, West Bengal, Maharashtra, Kerala and Gujarat, use over half of the are for agriculture.

The ratio of *total cropped area* to the *net area sown*, provides an index of cropping intensity. The agriculturally rich states and the hilly states have high cropping intensity. Kerala is an exception where the intensity is high due to the agricultural practices and population pressure on limited available area. In plain areas, the higher intensity is also due to irrigation facilities.

The Great Plains, Brahmaputra valley, coastal Gujarat, Andhra Pradesh, Orissa, Tamil Nadu and Kerala, interior Maharashtra and Karnataka, parts of Madhya Pradesh and Rajasthan, valleys in Himalayas constitute the chunck of arable land of the country. In addition there are small patches of arable land scattered all over. Here the population density is high and reasonable parts are irrigated. The forest cover the relatively rugged terrain of northeastern India, Himalayan region, Western Ghats, Godavari basin, parts of Madhya Pradesh, Bihar and Uttar Pradesh. There are small patches of forests all over the country, including the Ganga plain.

Wasteland

The area covered by scrub and grass is very limited. It is mostly in Gujarat and Rajasthan and in patches distributed everywhere. The wastelands are also found all over. Recently, with the help of several institutions, the National Wasteland Development Board, the wastelands of different types have been mapped throughout the country. The permanent unproductive land is found in Thar desert (Rajasthan), snow bound areas of Jammu and Kashmir, Uttar Pradesh and northern Sikkim.

Operational Holdings

The landuse at state level gives an aggregated pattern. At lower level, it is very complex due to fragmentation and size and ownership of holding of land. Private ownership of operation holdings allows decision-making about landuse at the lowest level. Past experiences, current prices, irrigation facilities, fertilisers and above all the estimation of monsoon conditions determine how the land is to be used in the forthcoming cropping season. The decision-making becomes more complex when other options, government sponsored or otherwise, are more lucrative, such as social forestry, industry, grazing, commerce, plantation and the like. These are extra-crop options.

The basis for understanding landuse at the village level is obviously the size of the operational holdings. In the first agricultural census of the country in 1970-71, information on this aspect was collected. Then, there were 70.5 million such holdings covering an area of 162 million ha. In the subsequent censuses, the number of holdings increased to 81.57 million in 1976-77 and 89.39 million in 1980-81, however the area remained almost same. Further, there has been a substantial increase (12.54%) in marginal (below 1 ha) and small (1-2 ha) landholdings, during the above two censuses. The numbers of other holdings have reduced. Now, about two-thirds of the holdings are below 2 ha. The average size is 1.82 ha (Table 10.3).

TABLE 10.3: Distribution of operational holdings, 1980-81.

Sub-groups	No. of operational holdings in million (%)	Area in million hectares (%)
Marginal (below 1 ha)	50.38 (56.6)	19.73 (12.1)
Small (1-2 ha)	16.10 (18.0)	23.01 (14.2)
Semi-medium (2-4 ha)	12.48 (14.0)	34.53 (21.2)
Medium (4-10 ha)	8.08 (9.0)	48.32 (29.7)
Large (10 ha)	2.15 (2.4)	37.17 (22.8)
Total	89.39 (100.00)	162.76 (100.00)

Source: Indian Agriculture in Brief.

The small size of holdings has operational limitations. Only selected drops can be economical which in most cases for consumption in the farmers' household. The commercialisation of such unit of land is near impossible. Obviously, the use of such holdings will be governed by the considerations which are not necessarily linked with the market econômy.

REFERENCES

Shafi, Mohammad (1984), *Agricultural Productivity and Regional Imbalances : A Study of Uttar Pradesh*, Concept, New Delhi.

NATMO (1978), *Atlas of Agricultural Resources of India*. National Atlas and Thematic Mapping Organisation, Calcutta.

Chatterjee, S.P. (1973), India, in *World Atlas of Agriculture*, Asia and Occania Volume, International Association of Agricultural Economists, Novara, pp. 186-200.

Directorate of Economics and Statistics (1986), *Indian Agriculture in Brief*, Ministry of Agriculture, New Delhi, 21st edition.

11

IRRIGATION

In India agriculture is the important form of economy. But this major occupation is greatly hampered by scanty rainfall in large areas and by erratic monsoon elsewhere. Most of the rainfall in India is due to the southwest monsoon. The average annual rainfall decreases to the west in northern India and suddenly decreases away from the coast of Western Ghats. Its arrival or departure fluctuates and it may fail in the middle of the season. Besides, there is a great variability from the normal which increases where the amount of rainfall is less. Partial failure or delayed arrival of the monsoon can cause extensive damage to the crops. Indian agriculture is considered as a gamble of monsoon. It is nearly so where irrigation does not exist. Hence, irrigation is necessary. The rapid growth of population indicates continuously increasing demand for food crops and other crops. Thus to maximise production of different cash crops and foodgrains, irrigation is inevitable.

Irrigation has proved beneficial to the country. In fact, irrigation forms the datum line for successful agriculture. The irrigation has provided a greater degree of in agricultural production system. It has certainly increased the yields.

Sources of Irrigation

Irrigation in India has been practised from ancient times and irrigation by tanks and wells are very familiar to the agrarian landscape of India. At the beginning of the Plan periods (1951), the area under irrigation was about 22.7 million ha. At present the total area under irrigation is about 67.5 million ha (Table 11.1). There are three principal types of irrigation, based on their sources : wells, tanks and canals (Map 11.1).

TABLE 11.1: Net area under irrigation and irrigation projects

State/UTs	Area under irrigation (%)	Canals	Tanks	Wells	Other Sources	Irrigation Projects up to VII Plan	
		(in percent)				Major	Medium
1	2	3	4	5	6	7	8
Andhra Pradesh	8.92	48.90	25.99	22.41	2.70	30	141
Arunachal Pradesh	0.06	—	—	—	100.00	—	9
Assam	1.47	63.28	—	—	36.72	3	22
Bihar	7.61	37.04	3.25	33.73	25.98	24	107
Goa, Daman & Diu	0.03	—	—	—	100.00	1	6
Gujarat	5.16	18.32	2.04	79.33	0.31	18	153
Haryana	5.50	54.40	—	45.31	0.29	14	18
Himachal Pradesh	0.23	2.17	—	2.17	95.66	1	6
Jammu & Kashmir	0.78	93.75	0.65	1.31	4.29	3	25
Karnataka	3.51	40.11	22.33	26.82	10.74	20	54
Kerala	0.61	44.11	23.53	—	32.36	13	14
Madhya Pradesh	6.01	44.34	5.83	42.32	7.51	30	142
Maharashtra	4.96	20.64	14.79	56.93	6.64	57	227
Manipur	0.17	—	—	—	100.00	4	3
Meghalaya	0.12	—	—	—	100.00	—	1

(Contd.)

(*Contd.*)

1	2	3	4	5	6	7	8
Mizoram	0.02	—	—	—	100.00	—	1
Nagaland	0.15	—	—	—	100.00	—	—
Orissa	3.13	65.92	17.04	17.04	—	11	51
Punjab	8.71	42.28	—	57.33	0.39	11	10
Rajasthan	7.69	31.54	3.75	62.82	1.89	7	106
Sikkim	0.25	—	—	—	100.00	—	—
Tamil Nadu	6.62	34.59	22.95	41.51	0.95	24	56
Tripura	0.07	—	—	—	100.00	—	3
Uttar Pradesh	24.16	33.62	1.75	61.35	3.28	48	95
West Bengal	3.84	64.47	20.34	1.14	14.05	10	23
UNION TERRITORY							
Andaman Nicobar	—	—	—	—	—	—	—
Dadra & Nagar Haveli	—	—	—	—	—	—	—
Delhi	0.13	30.19	—	58.49	11.32	—	1
Lakshadweep	—	—	—	—	—	—	—
Pondicherry	0.06	42.31	—	57.69	--	—	1

Source: Directorate of Economics and Statistics, Ministry of Agriculture, and Central Board of Irrigation Power.

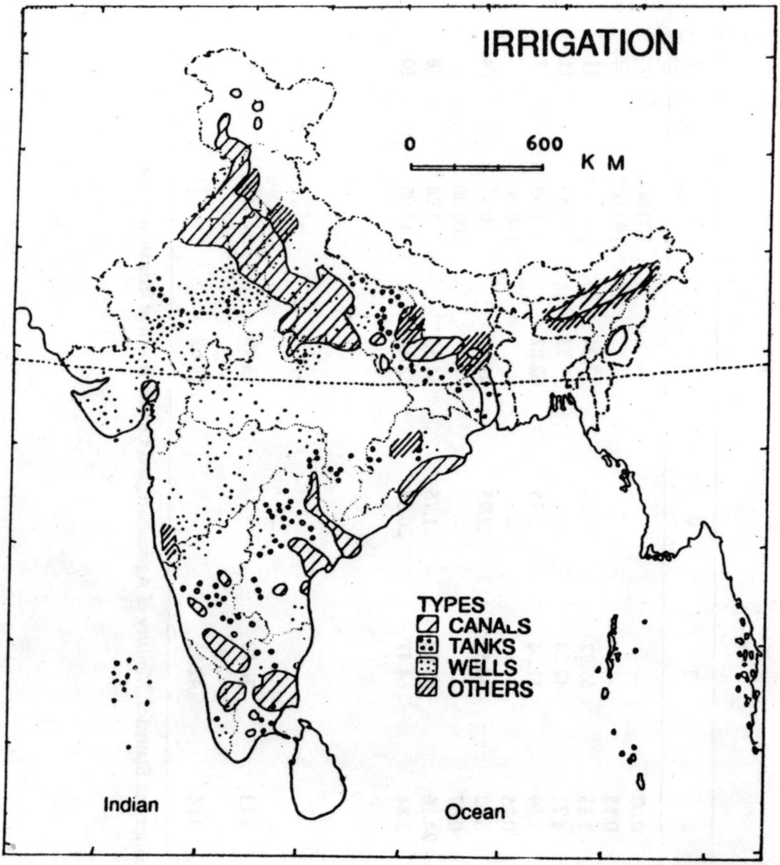

MAP 11.1: Irrigation

Wells

Well irrigation involves the lifting of water from the subsoil. The depth of the water in the subsoil varies widely in different parts with the variation in rainfall. About 11.8 million ha of land are irrigated by wells which is mainly confined to eastern and northern Uttar Pradesh, Bihar, eastern and southern parts of the Western Ghats including the states of Maharashtra, Karnataka, Madhya Pradesh, Tamil Nadu, Rajasthan and Gujarat. In India there are about 7 million wells. Water from the well are raised by four methods : (a) lever methods which are very common in Bihar and Andhra Pradesh, (b) inclined plane method, normally run by a pair of bullocks is mostly used in Uttar Pradesh, (c) Persian wheel mostly used in Punjab and western Rajasthan, and (d) tube wells which are very common in areas where the water-table is over 15 m deep. About 90 per cent of these tube-wells are located in Uttar Pradesh. Tube-wells are also found in Punjab, Haryana, Gujarat, Bihar and Madhya Pradesh.

Tanks

Tank irrigation is very common in the eastern part of Deccan plateau where the topography is suitable for erecting dams across small rivulets and for collecting water in artificial lakes so formed. Water is taken from the tanks running downward into the field. While the big tanks contain water all the year round, the smaller ones dry up in early summer. Tamil Nadu, Karnataka, Maharashtra, Andhra pradesh, West Bengal, Orissa, Uttar Pradesh and Bihar have large tanks and lakes which supply water for irrigation. In the multi-purpose river valley projects large tanks or reservoirs have been developed due to the construction of dams and barrages. About 4.3 million ha of land are irrigated by tanks.

Canals

Canals are by far the most important sources of water for irrigation. As the construction of canals is easy in the plains, the northern plains and the eastern coastal plains are the most important areas of canal irrigation. The canal may be either perennial or inundated. Perennial canals supply water throughout the year, while the inundated canals provide irrigation mostly during the rainy season.

The canals of the peninsula are taken out from the reservoirs built in the upper reaches of the river. The canals of the great plains are drawn from the rivers of the area, namely the Ganga, Yamuna, Ghaghra, and Gomati.

In undivided India, the most important canal systems were developed in Punjab. Among them the only Bari Doab canal lies partly in India. In Punjab other important canals are Western Yamuna canal, Sirhind canal, Bhakra canal, Nangal canal and the Gurgaon canal.

In Uttar Pradesh, Lower and Upper Eastern Yamuna, Agra, Sarda, Betwa, and Ramganga canals are important. Most of these canals are very old and need proper care. The canal irrigation has transformed certain areas of the deserted Rajasthan into a fertile state. There are numerous canal irrigation projects in Orissa, West Bengal, Bihar Gujarat and Maharashtra.

In the peninsula plateau, the Cauvery delta canal is said to be the oldest in India. In the Krishna delta the canals are constructed on both sides of the river from the barrage at Vijayawada town. In the Godavari delta, a dam in four parts is built across the two branches of river Gautami and Basistha. Besides these there are many delta projects in the peninsula. A number of small projects combining hydel power production and irrigation have been built in the Western Ghats at the headwaters of the rivers Godavari and Krishna and their tributaries.

Irrigation Potentials

All these above-mentioned sources of irrigation are grouped as major, medium and minor projects depending upon their Culturable Command Area. The irrigation potential, including major and medium irrigation projects, was 300 lakh ha in 1985-86; and minor irrigation, which includes surface water and ground water, was 375 lakh ha during the same period. The minor thrust of the development policy for irrigation sector is towards achieving the assessed target of 1,130 lakh of gross irrigation potential in the country based on conventional methods of diversion and storage by A.D. 2010 or so. Of these, about 585 lakh ha are excepted to be irrigated eventually from major and medium projects and balance by minor schemes (Appendix IV).

Major and Medium Irrigation

Irrigation projects with a Culturable Command Area (CCA) of more than 10 thousand ha are classified as major projects and projects with a CCA more than 2 thousand ha to 10 thousand ha are categorised as medium projects. Between 1951 and 1985, 246 major and 1,059 medium projects were taken up for execution. Among them 65 major and 626 medium projects were completed by 1985. During the Seventh Plan 18 new major and 66 new medium projects are being taken up. Out of the 199 major and 499 medium projects in hand, 58 major and 303 medium projects are expected to be completed during the Seventh Plan Period (Table 11.2).

TABLE 11.2: **Major and medium irrigation projects taken up and completed**

Plan	Major		Medium	
	Taken	Completed	Taken	Completed
I. (1951-56)	24	1	312	44
II. (1956-61)	23	4	192	112
III. (1961-66)	27	9	154	79
Annual Plans (1966-69)	11	1	32	58
IV. (1969-74)	33	9	66	72
V. (1974-78)	73	4	214	74
Annual Plans (1978-80)	14	1	28	25
VI. (1980-85)	41	36	57	152
VII. (1985-90)	18	58	66	303
Total	264	123	1125	819

Source: India 1988-89 and Central Water Commission.

Statewise distribution of major and medium irrigation projects shows that Maharashtra has completed highest number of both major and medium irrigation projects. It also has highest number of ongoing projects. It is mainly due to the fact that cotton and sugarcane, main crops of this state, require irrigated water. Major and medium irrigation facility is minimum in the northeastern states and Sikkim. Medium projects are large in number in drought prone areas like Rajasthan, Gujarat and Madhya Pradesh. Number of major and medium irrigation projects started during pre-independence period has been enlisted in Table 11.3.

TABLE 11.3: Completed major and medium irrigation projects

States/Union Territories	Major		Ongoing	Medium		Ongoing
	Pre-Plan	Plan Period	New of VII Plan	Pre-plan	Plan Period	New of VII Plan
Andhra Pradesh	11	5	14	45	41	45
Arunachal Pradesh	—	—	—	—	—	9
Assam	2	—	3	—	7	15
Bihar	2	6	16	2	66	141
Goa	—	—	1	—	—	3
Gujarat	—	7	11	—	71	82
Haryana	1	2	11	—	12	6
Himachal Pradesh	—	—	1	—	2	4
Jammu & Kashmir	1	1	1	6	7	18
Karnataka	4	2	14	12	23	19
Kerala	—	1	12	—	9	5
Madhya Pradesh	2	7	21	18	84	40
Maharashtra	5	10	42	16	125	86
Manipur	—	—	4	—	—	3
Meghalaya	—	—	—	—	—	1
Mizoram	—	—	—	—	—	1
Nagaland	—	—	—	—	—	—
Orissa	2	3	6	3	18	30
Punjab	3	6	3	—	7	3
Rajasthan	1	—	6	42	49	15
Sikkim	—	—	—	—	—	—
Tamil Nadu	16	5	3	7	34	15
Tripura	—	—	—	—	—	—
Uttar Pradesh	15	9	24	—	66	29
West Bengal	3	1	6	—	5	18
Union Territories						
Andaman & Nicobar	—	—	—	—	—	—
Chandigarh	—	—	—	—	—	—
Dadra & Nagar Haveli	—	—	—	—	—	—
Delhi	—	—	—	—	—	1
Daman & Diu	—	—	—	—	—	3
Lakshadeep	—	—	—	—	—	—
Pondichery	—	—	—	—	—	1
Indian	66	65	199	151	626	499

Source: Planning and Progressing Wing, Central Water Commission.

Minor Irrigation

The minor irrigation programme includes groundwater development by constructing dugwells, private shallow tube-wells, deep public tube-wells, boring and deepening of dugwells and small surface water development works through diversion schemes, storage schemes and lift irrigation, each having culturable command area of not more than 2 thousand ha.

Groundwater development, which constitutes bulk of minor irrigation programme is essentially a people's programme implemented primarily through individual and co-operative efforts with finance obtained mainly from institutional sources.To encourage water saving devices such as sprinklers, drip system, hydrants, water turbines and hand pumps. Government subsidies are made available to small and marginal farmers for their purchase.

Minor irrigation schemes provide instant and reliable source of irrigation to the cultivators. It also provides critical help in improving the status of irrigation and in controlling waterlogging and salinisation in the canal common area. In minor irrigation surface water projects, which are financed through Plans funds, often provide the means of irrigation in several undulating tracts comprising bulk of chronically drought affected areas (Table 11.4). Initial investment on these schemes is comparatively low and they can be completed quickly. Moreover, these schemes are labour intensive and offer employment to the rural people (Appendix V).

TABLE 11.4: Minor irrigation project (cumulative)

Period	Potential (Lakh hectares)
Pre Plan Period	129.00
First Plan	140.00
Second Plan	147.00
Third Plan	170.00
Annual Plan (1966-69)	190.00
Fourth Plan	235.00
Fifth Plan	273.00
Annual Plan (1978-80)	300.00
Sixth Plan	375.00
End of 1987-88	422.92
Ultimate potential (2010 A.D.)	550.00

Source : India, 1988-89.

Co-ordinated Efforts

The Government of India has also launched various programmes for the better utilisation of irrigation potential of the country. The command area development programme was launched at the beginning of Fifth Plan with the specific objective of ensuring a faster and better utilisation of irrigation potential in selected major and medium irrigation projects in the country. The programme broadly covers farm development works which include construction of field channels and field drains, land shaping wherever necessary, farm roads, consolidation of holdings and realignment of boundaries, introduction of warabandi or the rotational supply of water and setting up of WHF wireless network for better communication in order to ensure equitable and assured supply of irrigation water to each and every farm holdings, arrangements for supply of inputs and credits, agricultural extension, construction of markets and godowns and the development of ground-water for conjunctive use. The programme covering 131 selected major and medium irrigation projects in 20 states and two union territories with total culturable command area of over 180 lakh ha is executed through 50 command area development authorities.

REFERENCES

Kayastha, S.L. (1983), Floods in India: A study of their occurrence, causes, forecasting and control, *National Geographical Journal of India*, Vol. 29 (3 & 4), pp. 121-141.

Mulik, A. (1989), *Dynamics of Irrigation*, Bombay.

NATMO (1990), *Water Resources Development Atlas of India*, National Atlas & Thematic Mapping Organisation, Calcutta.

NATMO (1990), *Irrigation Atlas of India*, 2nd Ed. National Atlas & Thematic Mapping Organisation, Calcutta.

Rao, K.L. (1975), *India's Water Wealth*, Orient Longman, Calcutta.

12

AGRICULTURE, LIVESTOCK AND FISHERY

Land is India's major resource and about three quarters of the working population makes living out of it. Since Independence, the greatest effort in the exploitation of the country's resources have been directed towards the field of agriculture and allied sectors. Of the total 172 million ha of cropped land as much as 125 million ha is used for producing foodgrains, cereals and pulses. Rice occupies the foremost place. It is followed, in order of area occupied, by *jowar*, wheat, *bajra*, maize, barley and ragi (Table 12.1).

Foodgrains

Rice, the staple food of the majority of the Indian people is always preferred by the farmers as a field crop because its yields are highest among all the foodgrains. It grows best in the humid alluvial plains of West Bengal, middle Ganga plain and the Assam valley. Another extensive tract of paddy cultivation lies on the humid western coast. The permanent irrigated rice fields are located mainly in the upper Ganga plain and in the deltas of the eastern coast. The remaining rice fields lie scattered all over the country under various climatic conditions and at various altitudes, from sea level to about 2,500 m. In some hilly regions, such as Nagaland and Manipur, rice fields constitute more than ninety per cent of the total cropped land. Three crops of rice can be taken in the one year in the main rice growing areas, the main crop (winter or *aman* rice) being sown in June-July and harvested in November-December. The two other crops are autumn (*aus*) rice, sown in May-June and harvested in September-October and spring or summer rice, sown in November-December and harvested in March-April (Appendices VI and VII).

Jowar (Sorghum) is normally grown as a rainfed kharif crop (June to October and November), although it can be grown as a rabi crop

(October to February and March) in some parts of the country, especially on the moisture retentive deep black soils of Gujarat and the eastern parts of the Deccan, which receive rains during the retreating monsoon season under irrigation it can be grown .in both cropping seasons as grain and as green fodder. About eighty per cent of the area under wheat is located in the western part of the great plain in Punjab, Uttar Pradesh and in the central highlands in Madhya Pradesh. Two varieties are important: bread wheat, grown mostly in the western part of the great plain; and macaroni wheat, grown mostly in the black soils of the central highlands and northern Deccan.

Bajra is the most sturdy grain crop as it can withstand heat and drought and can thrive even on poorer soils. The dry sandy oils of Rajasthan have the largest area under this crop. Adjoining land of similar types in the Punjab and Gujarat is also suitable for *bajra.* The dry rain shadow areas of the Deccan states — Maharashtra, Karnataka, Andhra Pradesh and Tamil Nadu also have *bajra* growing areas. Like *jowar*, *bajra* is used widely as a forage crop especially in Punjab and south Deccan.

More than fifty per cent maize is produced in the lower Himalayan region, from Jammu and Kashmir eastwards through Uttar Pradesh, with a yield about the same as that of rice. In Bihar and Uttar Pradesh the crop is grown on fertile well drained loamy soils whereas in the mountainous tracts it can been grown on the coarser terrace soils. Maize is also used as a fodder crops (Fig. 12.1).

Pulses

The average annual production of pulses was some 13 million tons in 1985-86, the yield per ha being 7.6 quintals. The western part of the great plain, eastern Rajasthan and the central highlands have the largest area growing Bengal gram, half of the area under red gram is found in the upper and middle Ganga plains of Uttar Pradesh and Bihar and the central highlands of Madhya Pradesh. The remainder is concentrated in northern Deccan and in Karnataka, Andhra Pradesh and Tamil Nadu.

Oil Seeds

The chief industrial crops are oilseeds, cotton, jute, sugarcane and tobacco. A wide range of oil seeds such as groundnut, mustard, rapeseed, toria, taramia, sesame, lineseed and castor are produced. Others are niger, and sunflower. India has the largest area and the highest production of groundnuts in the world. About ninety per cent crop is grown in Gujarat and the four Deccan states : Maharashtra, Karnataka, Andhra Pradesh and Tamil Nadu. India also has the world's largest area and production of mustard and rapeseed, and of sesame. The first two oilseeds are concentrated in the upper Ganga plain, in Uttar Pradesh, Punjab and the Brahmaputra Valley, while sesame is found mainly in the upper Ganga plain, eastern Rajasthan, Andhra Pradesh and Tamil Nadu. The production of castor is much more localised than that of other oil seeds in Andhra Pradesh and Gujarat.

Fibre Crops etc.

The cotton is grown as a rainfed crop in (a) black cotton soils of the northern lava plateaus of Madhya Pradesh; (b) alluvial plain and Kathiawar lava plateau of Gujarat; (c) Vidarbha and upper Godavari plateau of Maharashtra and upper Krishna plateau of Karnataka in Telangana and the Cuddapah basin of Andhra Pradesh; and (d) Tamil Nadu uplands. In the Punjab plain and the upper Ganga plain of Uttar Pradesh it is grown mainly as irrigated crop.

Jute was the monopoly of India in the world markets until 1947. It thrives well in a humid and hot climate, especially on deep alluvial soils. Practically the entire production is in West Bengal and Bihar, Orissa, the Assam valley and Tripura.

The cultivation of sugarcane is confined mainly to those areas that have assured supplies of abundant irrigation water. The irrigated upper Ganga plain of Uttar Pradesh and the Punjab plain have the largest area under this crop. Sugarcane is also produced in north Sahyadri piedmont zone upper Godavari plateau, the upper Krishna plateau, Krishna-Godavari deltas, Visakhapatnam, coastal plain, northern Telengana and the southern part of the Cuddapah basin, the Cauvery delta, and parts of the Tamil Nadu upland around Coimbatore.

Plantation Crops

India is third largest producer of tobacco in the world, after the United States and China. This crop is grown mainly in the rabi season, without irrigation in the Krishna-Godavari delta. Tea is by far the most important plantation crop, most of the plantations being in the northeast — the Brahmaputra valley, east Meghna plains the piedmont plains of north Bengal, and on hill slopes in the Darjeeling Himalaya. It is grown on a smaller scale in the Kangra valley of Himachal

Pradesh and also in the Nilgiri mountains and around Coimbatore in the southern Sahyadri of Kerala, particularly in Kottayam district.

Coffee is confined to southern India, the ideal location being deforested areas with well drained rich loamy soils. Most of the crop come from Malnad region of Karnataka, the Nilgiri mountains, and the forest cleared areas of the southern Sahyadri. Tea and coffee plantations often occur side by side. Rubber is confined to the extreme southwest mainly in the Sahyadri hills and the Kanyakumari district of Tamil Nadu. The plantations are usually located on the lower hills slopes, tea and coffee being on the higher slopes.

Cardamom and pepper are the most important spices both being obtained from evergreen forests in the central and southern Sahyadri (Karnataka and Kerala). Turmeric, chillies and ginger are more widespread, both as garden and plantation crops.

Fruits and Orchards

Fruit trees, bushes and orchards land lie scattered throughout the country adjacent to the villages. Temperate fruits like apples, apricots, cherries, peaches and plums are grown in the Himalayan region. Bananas occupy the largest area under the indigenous fruit. Other tropical fruit are mango, litchi, guava, oranges, cashew, jack fruit, papaya and pineapples.

In India coconuts are used as fruits although they are also used for the manufacture of copra, from which oils is extracted. It is perennial tree grown as a permanent garden or plantation crop occurring mostly on the shores of the Malabar coast, mainly in Kerala. It represents the main crop of the Laccadive, Minicoy and Amindivi islands, and about half of the crop area reported in the Andaman and Nicobar islands. India has variety of agricultural products, so it is essential to divide India into proper agricultural regions.

Agricultural Regions

An agricultural region is defined as an having homogeneity in farming practices, crop produced and crop association. India is a vast country with great variations in the nature of farming practices and crop production (Table 12.2). The Indian Council for Agricultural Research has divided the country in agricultural regions by considering the production of crops. Based on the predominance of crops and crop associations, India can be divided into seven agricultural regions:

TABLE 12.1: Percentage share of principal crops to total cropped area, 1984-85.

State/Union Territories	Cereal & Millets	Pulses	Sugar-cane	Planta-tion crop	Fruits	Potato	Onion	Vege-table	Oil seed	Fibres	Tobacco
Andhra Pradesh	57.3	11.4	1.6	3.0	2.0	—	0.1	0.7	16.4	4.2	2.0
Assam	68.5	3.6	1.4	2.1	1.5	1.2	0.1	3.4	18.3	3.8	0.1
Bihar	76.7	12.3	1.3	0.4	1.3	1.4	0.1	3.2	2.5	1.6	0.1
Gujarat	39.2	7.1	0.9	1.3	0.7	0.1	0.1	0.8	25.2	14.8	1.1
Haryana	61.3	10.6	2.8	0.2	0.1	0.2	—	0.7	3.3	7.6	—
Himachal Pradesh	86.1	4.5	0.3	0.3	3.0	1.4	0.1	2.2	2.4	0.1	—
Jammu & Kashmir	80.5	4.9	0.1	0.2	3.9	0.2	—	1.5	5.6	0.1	—
Karnataka	52.0	14.3	1.7	2.5	0.9	0.1	0.3	1.0	14.9	8.6	0.4
Kerala	29.3	1.0	0.3	8.9	11.7	—	—	10.2	24.5	0.2	—
Madhya Pradesh	58.5	23.1	0.4	0.6	0.2	0.1	0.1	0.6	9.9	2.7	—
Maharashtra	55.3	12.9	1.6	1.0	0.6	0.1	0.3	0.7	10.1	13.7	0.1
Manipur	87.2	2.1	1.0	3.7	2.1	1.0	—	1.0	2.1	0.5	—
Meghalaya	63.4	1.4	—	7.3	1.4	8.7	—	13.0	3.8	9.1	0.5
Nagaland	80.3	2.8	2.8	1.7	3.4	3.4	—	5.1	2.2	—	—
Orissa	57.6	19.7	0.6	1.6	2.4	0.1	0.3	7.1	9.7	1.1	0.2
Punjab	69.5	3.0	1.5	0.1	0.3	0.4	—	0.9	2.7	10.5	—
Rajasthan	50.1	19.2	0.2	1.4	0.1	—	—	0.2	9.7	2.2	—
Sikkim	58.3	7.8	—	14.6	—	2.9	—	2.9	8.4	—	—
Tamil Nadu	56.9	8.2	3.1	3.0	2.6	0.1	0.3	1.8	16.4	3.2	0.1
Tripura	78.2	1.6	0.5	1.1	6.3	0.5	—	4.7	2.1	3.7	0.3
Uttar Pradesh	69.3	12.4	7.5	0.2	1.0	1.1	0.1	1.9	3.5	0.3	—
West Bengal	74.7	5.4	0.4	0.5	0.9	1.7	—	4.3	4.6	6.6	0.2
Others	71.4	1.7	0.9	0.7	7.9	0.5	—	2.5	10.1	0.8	0.2
India	59.4	13.2	2.0	1.3	1.1	0.4	0.1	1.7	9.9	5.2	0.3

Source: Directorate of Economics and Statistics, Ministry of Agriculture.

TABLE 12.2: Relative share of states in production of principal crops (1984-85).

(in %)

States	Food-grains	Sugar-cane	Total oil seeds	Cotton	Jute
Andhra Pradesh	7.6	6.0	12.7	11.7	—
Assam	1.9	1.2	1.2	—	13.9
Bihar	6.4	2.3	1.1	—	13.6
Gujarat	3.6	4.1	17.8	22.6	—
Haryana	4.8	3.1	1.7	9.1	—
Himachal Pradesh	0.7	—	—	—	—
Jammu & Kashmir	0.8	—	0.5	—	—
Karnataka	4.6	7.8	8.3	9.0	—
Kerala	0.9	0.5	0.1	0.1	—
Madhya Pradesh	9.7	0.9	9.4	3.6	—
Maharashtra	7.0	15.7	11.2	17.4	—
Manipur	0.2	—	—	—	—
Meghalaya	0.1	—	—	0.1	0.7
Nagaland	0.1	0.1	—	—	—
Orissa	4.2	1.8	5.7	—	6.4
Punjab	10.5	3.1	1.2	14.1	—
Rajasthan	6.1	0.8	7.7	7.0	—
Sikkim	0.1	—	0.1	—	—
Tamil Nadu	4.2	9.0	9.2	4.9	—
Tripura	0.3	—	—	—	0.4
Uttar Pradesh	20.1	42.9	10.3	0.3	1.2
West Bengal	5.7	0.6	1.7	—	63.8
Others	0.4	0.1	0.1	0.1	—
India	100	100	100	100	100

Source: Directorate of Economics and Statistics, Ministry of Agriculture.

Fruits and Vegetables Regions

This region extends from Kashmir valley to Assam. The rainfall varies 2,000 mm in the east to less than 600 mm in the extreme northwest. In the western Himalayas most of the temperate fruits like apple, peach, cherries, plum, apricot, etc. are grown. In eastern Himalayas, oranges are important. Rice, maize, ragi, potatoes, chillies and vegetables are produced as well.

Rice-Jute-Tea Region

This region includes Assam, Arunachal Pradesh, Tripura, Meghalaya, West Bengal, Orissa, northern and eastern Bihar, terai region of Uttar Pradesh. It has vast areas of lowlands, river deltas, valleys which contain thick alluvial deposits. There is a considerable variation in the rainfall from 1,870 mm to 2,500 mm. Most of the rain comes during the southwest monsoon season. Fertile alluvial soils, abundant rainfall and high summer temperatures make rice predominant crop of the region.

In northern Bihar, rice is the most important crop and the other crops grown are sugarcane, jute and tobacco. The coastal regions, both east and west, a part from rice, coconut and mangoes are important products.

Tea plantation is important in upper Assam valley, Tripura and northern Bengal. Jute cultivation has gained great importance in Assam valley, Ganga delta and plains and in the east coast. Mustard is grown throughout the area. Mango, pineapple, betel leaves, bananas, jack-fruits, coconuts and oranges are the principal fruits.

Wheat and Sugarcane Region

This region comprises northern Bihar, Uttar Pradesh, Punjab, Haryana, western Madhya Pradesh and north eastern Rajasthan. The soil is mostly alluvial though there are patches of black and red soils. Rainfall is moderate and occurs both in winter and south west monsoon period. Irrigation plays a vital role in the drier regions. The most important crops are wheat and sugarcane. Rice is also cultivated. In Uttar Pradesh and in Bihar plains, north of Ganga, wheat and sugarcane are outstanding crops. The other crops grown are maize and pulses. The major wheat growing belt of India lies in the canal irrigated Ganga-Yamuna *doab* and in the plains of Punjab, Haryana and northern Rajasthan. The whole region has diversified agricultural crops.

Millets and Oilseeds Region

This region includes Karnataka plateau, parts of Tamil Nadu southern Andhra Pradesh and eastern Kerala. This region has poor soils, broken topography and less rainfall which varies from 750-1,250 mm. Millets grown include *bajra, ragi,* and *jowar,* while oil seeds include groundnut, castor and pulses. Mangoes, bananas are largely grown.

Maize and Coarse Crop Region

The region includes semi-arid and arid lands of western Rajasthan and northern Gujarat. The soil varies from alluvium in the east to sandy in the west. The rainfall is mostly below 500 mm. Agriculture is practised only when irrigation can provide water to the ground. The Mewar plateau is mainly a maize growing region but it also produces wheat and ragi. In the southern part rice, cotton, sugarcane are grown. *Bajra*, pulses are cultivated throughout the region.

Cotton Region

The region mostly includes the Deccan Trap, where black cotton soils cover major parts of the river valleys. The region is a rainshadow area with an annual rainfall varying from 750-1,000 mm. This is the main cotton belt of India. *Jowar, bajra*, gram, sugarcane, wheat, etc. are also grown in many parts of the region.

Spices and Plantation Crop Region

It includes both eastern and western coastal plains, Andaman and Nicobar islands, Lakshadweep and Minicoy islands. In this region rainfall is heavy and is over 2,500 mm. Plantation crops like coffee, rubber, tapioca, pepper, cardamom are three main crops. Rice is an important food crop. Due to the pressure of population in India, food crop important agricultural products all over the country. But animal husbandry also occupies an important place in the economy of India. No other country of the world has as many livestock as India. According to livestock census of 1982, it had 19.24 crore cattle and 6.98 crore buffaloes constituting about one-sixth and one-half of the world cattle and buffaloes population respectively.

Livestock

In India the need to organise a cattle development programme on intensive lines like linked dairy schemes has been recognised. The governments recently initiated cattle breeding tracts, upgrading of non-descript drought herds with better dairy breeds, cross breeding with exotic breeds in hilly areas and other regions that have facilities for rearing and maintaining high-yielding milch animals. Thirty intensive cattle development programmes are proposed. The greatest increase in numbers of livestock have been amongst cattle buffaloes and poultry, while the number of horses has been declining (Table 12.3).

In the country the production of milk is less compared to the number of cow which indicates that the dairy industry in India is poorly developed. The low yield of cow milk is ascribed to the many handicaps of the dairy industry, such as inadequate feed and fodder resources, the limited grazing land, and the lack of better breeds.

The buffaloes though much smaller in number produced more milk. Goats are principally used for meat and hides but sheep are used both for meat and wool. Most wool yielding sheep are raised in the arid plains of Rajasthan, Gujarat and Haryana and on the Himalayan meadows. The rapid rise in the number of ducks and chickens in relatively recent years is due to the starting of commercial chicken farms. Silkworm rearing is also practised in the states of Karnataka, West Bengal, Assam and Jammu and Kashmir.

Fishery

In some parts of India, fish serve as an important item of food. But the fish catches fall far short of the country's requirements. India's annual production of fish amounted to 25,40,700 tonnes of which 9 lakh tonnes is fresh water fish (1983). Though the production of marine fish has increased considerably in past 30 years but India with long coast line and wide continental shelf, can steep up the fish production (Table 12.4).

Maharashtra state with a 700 km of the Arabian sea coast is in the forefront of fishery development. Karnataka on the same coast with a 298 km long coast line and impounded water tanks and long stretches of rivers have been steadily increasing the production of fish both form inland and marine fisheries. In Goa, estuarine fisheries are also well developed. On the 100 km long Arabian Sea coast which is used for fishing, produces prawns, an important foreign exchange earner. However, the bulk of the marine fish production constitutes mackerel, sardines, verem, caranzalen. Dona Paula and Chiclaim are ideally located for developing them as fishing harbours in Goa. Minicoy and Lakshadweep islands have rich tuna fishery along their coasts. the Andaman Sea is also potentially rich in herring, salmon and sardines which are mainly caught around the coasts of the Andaman and Nicobar Islands and marketed at Port Blair. Kerala in the west coast has the largest marine sea fisheries, contributing to about one-fifth of the fish production of India. Quilon is the major fishing harbour on this coat. Estuarine fisheries are well developed a Zuari and Mandevi estuaries of Goa and some of the fish catches (smeli fish) are initially brought to the Mormugao harbour. The potential fisheries in the Bay of Bengal cover an area of about one lakh km² of which Sunderban estuaries are rich in many varieties of fishes, such as gurjali, pomfrets. A few these catches find their way to the large inland consumer centres

TABLE 12.3: Livestock and poultry, 1981

(*in percentage*)

States/ Union Territories	Total Live stock	Cattles	Buffaloes	Sheep	Goats	Other	Total Poultry
Andhra Pradesh	8.5	38.26	22.76	22.44	13.86	2.68	13.43
Arunachal Pradesh	0.2	28.81	2.06	3.43	13.03	52.67	0.47
Assam	2.6	68.91	7.62	0.61	17.29	5.57	6.49
Bihar	8.6	47.74	14.05	3.62	31.25	3.34	8.75
Goa, Daman & Diu	0.1	46.60	15.27	0.38	8.01	29.74	0.26
Gujarat	3.9	41.69	24.11	11.05	21.41	1.74	2.13
Haryana	1.9	35.37	42.58	7.85	7.52	6.68	0.87
Himachal Pradesh	1.3	43.92	5.48	22.00	27.84	0.76	0.20
Jammu & Kashmir	1.3	45.89	10.73	26.10	14.85	2.43	1.27
Karnataka	5.9	46.89	15.03	20.81	15.54	1.73	6.02
Kerala	1.4	56.51	8.53	0.05	31.64	3.27	8.32
Madhya Pradesh	10.9	65.09	14.49	2.40	16.67	1.35	4.44
Maharashtra	8.0	51.33	13.15	8.89	25.51	1.12	11.65
Manipur	0.1	57.99	10.25	0.39	3.15	28.22	0.58
Meghalaya	0.2	58.67	4.92	2.46	14.63	19.32	0.66
Mizoram	0.1	39.83	1.62	0.81	18.70	39.04	0.70
Nagaland	0.1	24.80	2.13	—	6.40	66.67	0.44
Orissa	5.0	65.07	7.29	7.68	18.34	1.62	5.90
Punjab	2.4	36.81	45.68	5.53	8.02	3.96	3.44
Rajasthan	11.2	31.18	12.26	24.03	29.75	2.78	0.9
Sikkim	0.1	54.11	1.71	5.48	30.48	8.22	0.14
Tamil Nadu	6.5	44.73	12.74	21.91	17.40	3.22	8.92
Tripura	0.2	69.40	1.64	0.35	23.33	5.28	0.41
Uttar Pradesh	13.8	49.24	26.68	3.93	16.17	3.98	3.42
West Bengal	5.1	62.24	4.31	4.15	27.30	2.00	9.69
Union Territories							
Andaman and Nicobar Islands	0.1	35.52	13.15	--	23.68	27.65	0.11
Chandigarh	0.1	15.79	63.16	5.26	10.52	5.27	0.03
Dadra & Nagar Haveli	0.1	69.09	5.45	—	23.64	1.82	0.03
Delhi	0.1	23.78	52.91	4.36	9.71	9.24	0.13
Lakshadweep	0.1	16.67	—	—	83.33	—	0.02
Pondicherry	0.01	62.16	6.75	3.37	26.35	1.37	0.09

Source: Directorate of Economics and Statistics, Ministry of Agriculture.

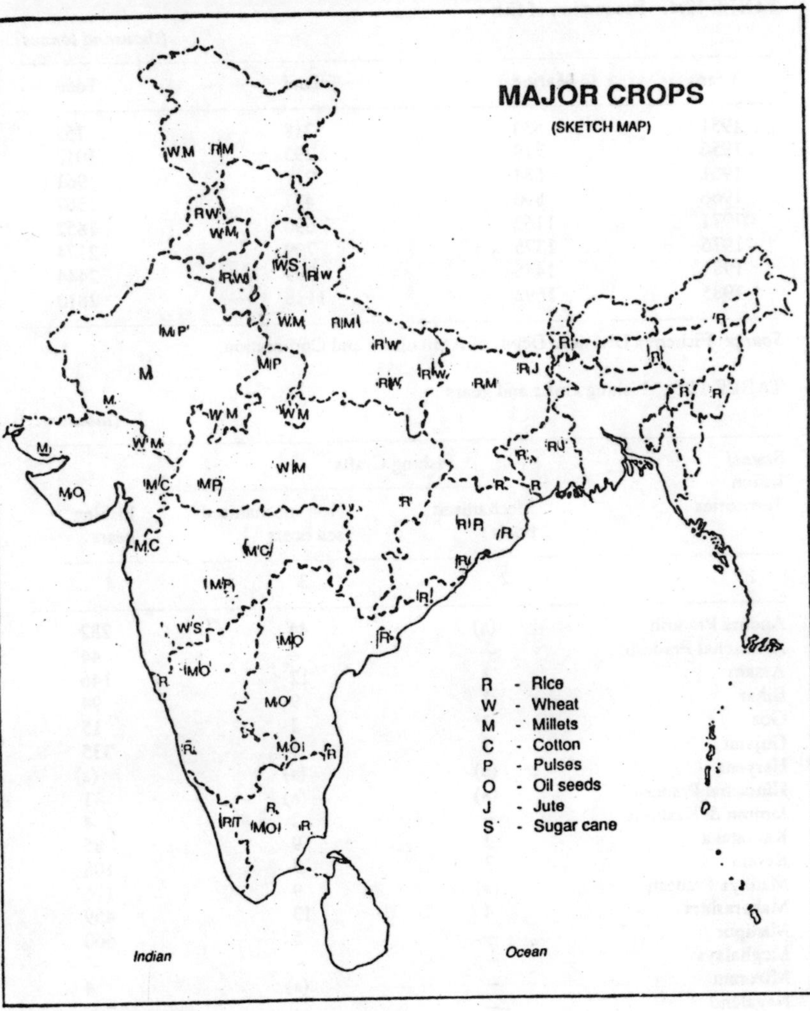

FIG. 12.1: Major Crops

due to lack of transport and preservation arrangements. Further, inland prawns, bhangar and crabs like *katla, rui, mrigel* and *bhetki* are caught and supplied to Calcutta. On the coast the Chilka Lake in Orissa is an important for fishing. It supplies fish to Calcutta as well. The country has the potentialities for having higher production if the infrastructure is further developed (Table 12.5).

TABLE 12.4: Production of fish

(thousand tonnes)

Year	Marine	Inland	Total
1951	534	218	752
1956	719	293	1012
1961	684	277	961
1966	890	477	1367
1971	1162	690	1852
1976	1375	799	2174
1981	1445	999	2444
1985	1692	1118	2810

Source: Fisheries Division, Deptt. of Agriculture and Corporation.

TABLE 12.5: Fishing crafts and gears

(thousands)

States/ Union Territories	Fishing Crafts		Fishing gears
	Mechanised boats	Non-mechanised sea boats	
1	2	3	4
Andhra Pradesh	(a)	44	282
Arunachal Pradesh	—	—	44
Assam	1	12	146
Bihar	1	9	94
Goa	1	1	15
Gujarat	3	5	335
Haryana	(a)	(a)	(a)
Himachal Pradesh	(a)	(a)	1
Jammu & Kashmir	—	2	4
Karnataka	1	9	85
Kerala	3	3	108
Madhya Pradesh	(a)	9	176
Maharashtra	4	13	459
Manipur	—	5	660
Meghalaya	—	—	—
Mizoram	—	(a)	4
Nagaland	—	—	—
Orissa	2	5	154
Punjab	—	—	—
Rajasthan	(a)	(a)	2
Sikkim	—	—	—
Tamilnadu	4	34	150
Tripura	—	3	13
Uttar Pradesh	(a)	11	91
West Bengal	(a)	18	1020

(Contd.)

(Contd.)

1	2	3	4
Union Territories			
Andaman & Nicobar Islands	(a)	1	2
Chandigarh	—	—	—
Dadra & Nagar Haveli	—	—	—
Delhi	—	(a)	1
Daman & Diu	—	2	17
Lakshadeep	(a)	1	1
Pondicherry	(a)	2	8
India	20	220	3,872

(a) Below 500
Source: Fisheries Division, Deptt. of Agriculture.

India has abundant potentially for the development of agricultural products and also livestock and fisheries. But, proper scientific approach is required. Though the production and consumption of fertiliser has increased manifold but the production of agriculture is not that satisfactory as it was in the later part of 1970s. Over utilisation of chemical fertiliser is causing soil degradation and it is unfortunate that Indian farmers still have to depend on monsoon for crops. The Government of India has paid little attention to the livestock development in the initial plan period, so the quality of livestock in this country is relatively poor.

REFERENCES

Bayliss Smith, Tim P. & Suddhir Wanmali (eds.) (1984), *Understanding Green Revolution*, Cambridge University Press, Cambridge.

Lal, M. (1986), *Agro-types in Central India*, Star Distributors (Publication Division), Varanasi.

Mohammad, A (1978), *Studies in Agricultural Geography*, Rajesh Publication, New Delhi.

NATMO (1976), *An Atlas of Agricultural Resources of India*, National Atlas & Thematic Mapping Organisation, Calcutta.

Reddy, N.B.K. (ed.) (1974), *Drought Prone Areas of India*, Rayalaseema Geographical Society, Tirupati.

Shafi, M. (1984), *Agricultural Productivity and Regional Imbalances: A study of Uttar Pradesh*, Concept Publishing Company, New Delhi.

Singh, J. (1974), *An Agricultural Atlas of India: A Geographical Analysis*, Vishal Publications, Kurukshetra.

Singh, V.R. (1974), Agricultural typology of India, *Geographica Polonica*, Vol. 40, pp. 113-32.

Srivastava, U.K. & M. Dharma (eds.) (1985), *Fisheries Development in India: Some Aspects of Policy Management*, Concept Publishing Co. New Delhi.

13

DISTRIBUTION AND GROWTH OF POPULATION

India has a great diversity from region to region in geographical, social and economic conditions. The impact of these socio-economic differences on the process of the distribution and growth of population is very important. Therefore it is important to study the variations in trends in population distribution and growth.

Distribution of Population

India covers an area of about 3.28 million km^2 and her population in 1981 was 684 million. The density was about 216 persons per km^2 (Table 13.1). The vast volume of aggregate population of India reveals hardly any relation between area and size of population. Thus, the fourth largest state, Uttar Pradesh has maximum population size where as the thirteenth state in area, i.e. West Bengal ranks fourth in the population size. The simple statewise density of population gives as very generalised idea of the density pattern of India. A detailed study of the density pattern of India by districts will throw more light on the factors which causes variation in the density pattern (Map 13.1).

There are virtually demographic deserts in some areas whereas density clusters elsewhere need explanation. Demographic deserts are, however, few and restricted to desolate mountainous regions, arid parts of Rajasthan and some remote interiors of Ladakh in Kashmir, Lahul Spiti and Kinnaur in Himachal Pradesh, Mizo hills, Mikir, Cachar, and Naga hills in the northeastern region, and Andaman, Nicobar and Laccadive islands. There are few other demographic semi-deserts in the neighbouring areas of the areas mentioned above. An analysis of factors influencing variation in density may be summarised in the following way:

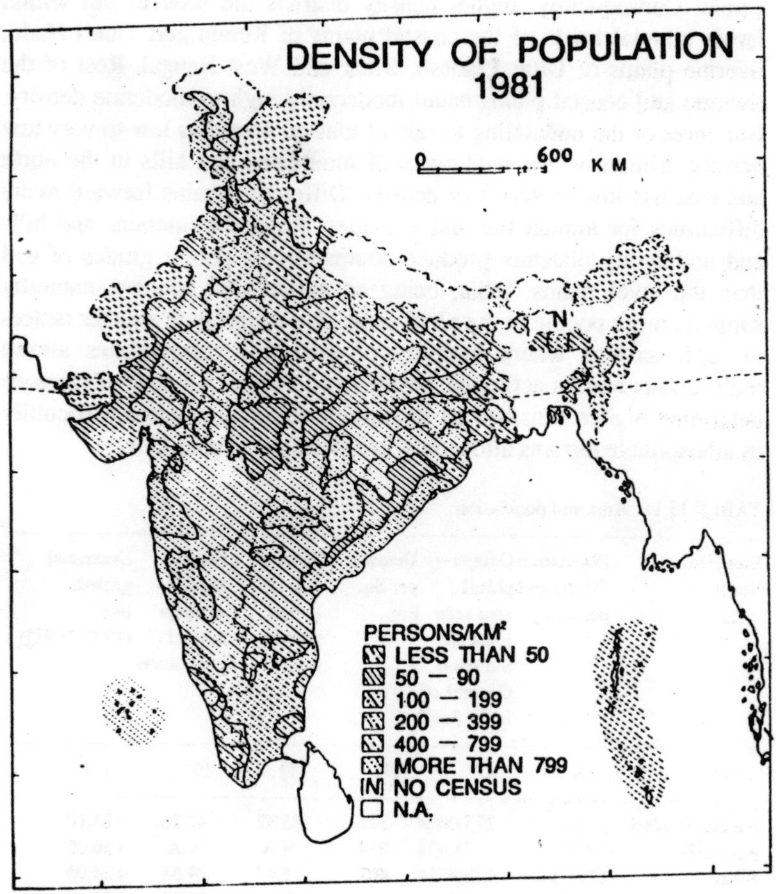

MAP 13.1: Density of Population

Influence of Terrain

Ease of living in comparatively level areas like riverine plains and basins, and coastal plains naturally encourage higher density as a whole. Consequently, higher density districts are seen to fall within fertile alluvial tracts of the coastal plains of Kerala and Tamil Nadu, riverine plains of Uttar Pradesh, Bihar and West Bengal. Rest of the riverine and coastal plains entail moderately high to moderate density. But, most of the undulating terrain of plateau areas has low to very low density. Similarly, the wilderness of mountains and hills in the north and east has low to very low density. Different terrains forward many difficulties for human life and activities. Rugged mountains and hills and undulating plateaux produce comparatively poorer grades of soil than the level plains. India, being an agricultural country, naturally supports more people in the plain areas. Accessibility is another factors in such terrains where mobility of men and commodities allows multifarious human activities, besides agriculture. All such advantages determine higher densities on favourable terrains, whereas difficulties in inhospitable terrains allow only low densities (Table 13.1).

TABLE 13.1: Area and population

State/Union Territory	Population (Thousand persons)	Geographical area according to Surveyor General of India (km^2)	Density per Sq. km.	Percentage of Urban population	Percentage of workers to total population	Decennial growth rate (1971-1981)
(1)	(2)	(3)	(4)	(5)	(6)	(7)
Andhra Pradesh	53,550	275,068#	195	23.32	42.26.	+23.10
AssamΦ	19,897	78,438	254	N.A.	N.A.	+36.05
Bihar	69,915	173,877#	402	12.47	29.68	+24.06
Gujarat	34,086	196,024#	174	31.10	32.22	+27.67
Haryana	12,922	44,212#	292	21.88	28.35	+28.75
Himachal Pradesh	4,281	55,673	77	7.61	34.36	+23.71
Jammu & Kashmir^^@	5,987	222,236#*59	178	21.05	30.76	+23.71
Karnataka	37,236	191,791	194	28.89	36.76	+26.75
Kerala	25,454	38,863#	655	18..74	26.68	+19.24

(Contd.)

(Contd.)

(1)	(2)	(3)	(4)	(5)	(6)	(7)
Madhya Pradesh	52,179	443,446#	118	20.29	38.41	+25.27
Maharashtra	62,784	307,690#	204	35.03	38.71	+24.54
Manipur	1,421	22,327	64	26.42	40.35	+32.04
Meghalaya	1,336	22,429#	60	18.07	43.44	+32.04
Nagaland	775	16,579	47	15.52	47.53	+50.05
Orissa	26,370	155,707	169	11.79	32.75	+20.17
Punjab	16,789	50,362	333	27.68	29.35	+23.89
Rajasthan	34,262	342,239	100	21.05	30.48	+32.97
Sikkim	316	7,096	45	16.15	46.60	+50.77
Tamil Nadu	48,408	130,058#	372	32.95	39.30	+17.50
Tripura	2,053	10,486	196	10.99	29.64	+31.92
Uttar Pradesh	110,862	294,411#	377	17.95	20.22	+25.49
West Bengal	54,581	88,752#	615	26.47	28.26	+23.17
A. & N. Island	189	8,249	23	26.30	33.21	+63.93
Arunachal Pradesh	632	88,743#	8	6.56	49.61	+35.15
Chandigarh	451	114	3,981	93.63	34.69	+75.55
Dadra & Nagar Haveli	104	491	211	6.67	40.81	+39.78
Delhi	6,220	1,483	4,194	92.73	31.93	+53.00
Goa, Daman & Diu	1,087	3,814	285	32.37	30.59	+26.69
Lakshadweep	40	32	1,258	46.28	19.74	+25.53
Mizoram	494	21,081	23	24.67	41.73	+48.55
Pondicherry	604	492	1,229	52.28	28.66	+28.15
All-India	685,185	3,287,263	2161	23.70	33.45	+26.00

Φ Projected figures for 1981.

\# Provisional.

^^ Area figures include the area under unlawful occupation of Pakistan and China where the census could not be taken.

* Includes area under illegal occupation of China and Pakistan.

@ Population figures exclude population of areas under unlawful occupation of Pakistan and China where census could not be held.

^ The density has been worked out on comparable data.

N.A. Not Available

Note: The total area of the country represents provisional 'Geographical Area' as on 31st March, 1982 supplied by the Survey of India.

Source: Census of India, 1981, Series—I, India, Part II B (i) Primary Census Abstract, General Population, 1981 Census.

Influence of Climate

Climate plays and important role in determining the density patterns. The twin elements of precipitation and temperature excercise major influence in such determination. India is still predominantly an agricultural country. Indian farmers continue to depend on annual rainfall for good harvest. As such, adequate and dependable amount of rainfall support higher density of population in Kerala, West Bengal and northern Bihar. However, there are exceptions. All high rainfall areas do not support high densities due to other physical factors which exist there. Assam valley, for example, has a moderate density pattern in spite of the fact that there is less variability in rainfall. Malabar and Circars coasts also are such areas of moderate densities with dependable annual rainfall. Similarly, the southern face of the Himalayas, though receives high rainfall has less density because of destructive floods, infertile soils and dense forests in the foothills zones. Mountain soils and high altitude together resist high density in those areas.

In the savana regions undependableness of rainfall is a common. In these regions the density pattern is uniformly low. Region with arid and semi-arid type of rainfall have very low to low density. Rajasthan and parts of Gujarat exhibit this type of density.

Temperature, throughout India is generally high which is not really detrimental to high density. Density of India is therefore, little effected by temperature excepting in arid and mountainous zones. Western Thar is one such area where density pattern is affected by high temperature besides low rainfall. On the other hand, low temperature also interferes with high density. On high altitude, beyond 2,000 m, life is intolerable. The coldness curb normal human activities and as a result density becomes very low.

Influence of Water Bodies

Absence of water is an insurmountable obstacle for the distribution of population in any form. In this country rivers are the most important waterbodies which besides rainfall greatly shape the density of population. Over the northern plains, large snowfed rivers sustain higher densities of population than most of the other regions of the country. Over the plains, comparatively drier regions like the Indo-Gangetic *doab* has lesser density than even some of the western regions having similar rainfall. The gift of the rivers in India is manifold. Its prime role is in the field of agricultural activity. It is essentially an agricultural country which depends largely on river waters for the purpose of irrigation. Transport and trade are the other factors which aid the cause of higher densities along the river.

Significant lakes are few in India. Yet some of the lakes in Rajasthan and Kashmir contribute commendably towards higher densities. It is also interesting to note that about 250 km of backwaters in Kerala have not reduced the general density of the region.

Influence of Soil

For an overwhelmingly agricultural country like India, the role of soil in shaping density of population is of great significance. Alluvial soil particularly rich with annual floods mostly supports high density of population. Other recent riverine soils are also equally contributory to their fertility. Black soils and podzol soils, though less fertile than the alluvial soils, but with irrigation can support a higher population density. However, certain soils like desert soil, mountain soil and saline soil are quite incapable of supporting high densities. Nevertheless, the existing pattern may change to a certain extent in future with new techniques of manuring and extension of major and minor irrigation works.

Impact of Minerals and Industries

When agriculture reaches saturation level under certain limiting conditions, opening of mines and industries play an important role in offering additional opportunities and thus promoting higher density of population. Besides agriculture, industrial activities have increased the density of population in southern West Bengal, Uttar Pradesh and in other places (Chapter 19).

Some industrial and mining regions do not show remarkable density due to physical obstacle. Mineral exploitation in South Bihar and Orissa and industrial activities of Jamshedpur have influenced the general densities of the region on a small scale. Similar results are observed in the southern plateau and in southwestern India. Industrial development also draws rural population to industrial spheres. As such, the balance of the overall density sometimes remains the same in a way by rural out-migration and urban industrial in-migration. Optimum dvelopment of both agriculture and industries can redistribute the lopsided density pattern of the country in future.

Impact of Urbanisation

Impact of urbanisation on population density is immense. In India urban centres may be classified into several categories according to functional characteristics. The various human activities have differential influences on density. Large towns naturally swells the population density of the concerned area or even more. In West Bengal, the industrial towns have definitely cater higher densities in the districts of Howrah Hugli, 24-Parganas and Barddhaman. Administrative towns like New Delhi, transportation towns like Mughal Sarai and Kharagpur, tourist centres like Srinagar, Darjeeling, and religious cities like Varanasi play similar roles in promoting higher densities in the areas concerned.

The comparable of density pattern in India though brings out a very uneven picture, it can hardly shed any light on extremities in densities. Neither does it give any indication regarding optimum density. An analysis of the character of growth of population may help in analysing of these aspects of population densities in India.

Growth of Population

In India the growth of population was sporadic during 1891-1921. Besides this irregularity, the growth rate of population, before 1921 was very low. The situation after 1921 is totally different. There was a rapid and uninterrupted growth (Table 13.2). Despite the World War II and the famine that occurred during the war, and the riots during the partition of the country in the decade 1941-51, the country witnessed a huge increase in population.

This growth of population is not due to any sudden burst in the birth rate but due to the decline in the death rate. The death rate started declining at a comparatively faster rate, while the birth rate remained stationary or declined slightly from 1921 onwards.

In order to understand the causes of population growth in India, it is essential to look into the changes in birth and death rates. It is largely the difference between these two rates which influences the growth rate of population of a country. Migration can play some role as well. But, in India, migration has no significant role except in some states.

The population growth rate was about 25 per cent during 1971-81 decade. In actual terms, there was an addition of 137.13 million population. The districtwise analysis indicates high growth rate in western Rajasthan, northeastern states etc. Parts of the Great Plains, Andhra coast, interior Maharashtra and northern India recorded growth rate closer to the country's average (Map 13.2).

TABLE 13.2: Population growth, 1891-1991

Year	Population in Millions	Decadal Growth	
		Absolute growth in Million	Percentage Growth
1891	239.00	—	—
1901	238.39	-0.61	-0.25
1911	252.09	+13.70	+5.75
1921	251.32	-0.77	-0.31
1931	278.97	+27.65	+11.00
1941	318.66	+39.64	+14.22
1951	361.08	+42.42	+13.31
1961	439.23	+78.15	+21.51
1971	548.15	+108.92	+24.80
1981#	685.18	+137.13	+25.00
1991@	845.00	+159.82	+23.33

Includes projected population of Assam.
@ Estimated.

Source: R.C. Chandra (1989), *Fertility: The most critical issue in Indian Demography, unpublished* paper.

At present, the birth rate is 32 per 1,000 people and the death rate is 12 per 1,000. The infant mortality rate is 114 per 1,000 (Appendix VIII). The pyramid of population in terms of age projects a tremendous increase over the present 730 million in the coming years. However, population growth in this century varies among states. In addition, the increase of urban population is much faster as compared to the rural population. Among the states like Maharashtra, Gujarat, Tamil Nadu, Punjab and West Bengal have shown faster rate of urban growth (Appendix IX).

To check the population growth it is essential to control net productive rate as soon as possible. In India the present level of fertility is high as compared not only with the developed countries, but also with many developing countries. There are various factors which are associated with fertility, such as the age of women at marriage, socio-economic characteristics including occupational status, educational status and the like.

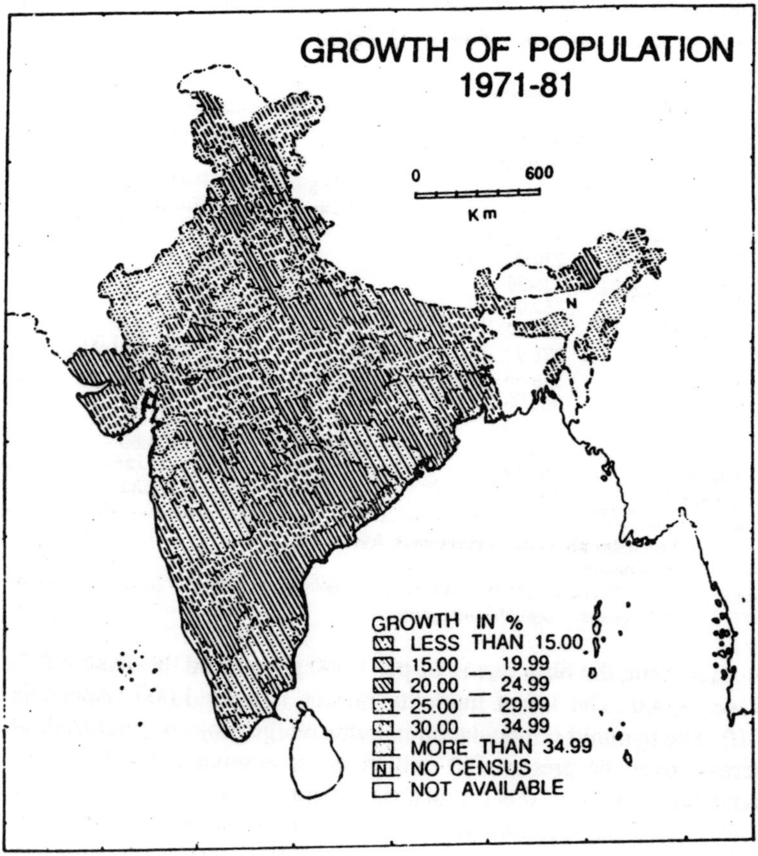

MAP 13.2: Growth of Population

There is variation among the states in all measures of fertility. At present, Uttar Pradesh has the highest fertility level and Kerala has the lowest level of fertility. An interesting finding is that the decline in fertility is faster in the states where the fertility level was higher in the earlier decades.

The high growth rate in India is due to the decrease of mortality rate. The mortality rate has decreased 66 per cent in this century. The present level of mortality is low when compared with India's past mortality levels but is high compared to the mortality level presently prevailing in developed countries. Therefore, there is further scope for a mortality decline. This is also substantiated by rural-urban differentials in mortality while the levels of mortality in urban areas are low, the levels of mortality in rural areas are high. The highest level of mortality was in the decade 1911-20. The period covers the First World War, and is known to be a decade of famines. The 1918 influenza epidemic also falls in this decade. The control of diseases and its prevention through the health care programmes and the availability of medical facilities, particularly after Independence, are responsible for the declining mortality to a great extent.

Internal migration within the country has some influence the growth of population in some areas of the country. Assam, West Bengal, Maharashtra, Gujarat and Karnataka are predominantly net in-migrating states. Bihar, Uttar Pradesh, Tamil Nadu and Orissa have been predominantly net out-migrating states. Kerala, Madhya Pradesh, Andhra Pradesh and Punjab showed fluctuating trends. The volume of migration streams was quite significant compared to the population either in the states of origin or in the states of destination.

There are two factors which cause migration. First, economic and second, political. The growth of tea estates in Assam, industrial and urban development in West Bengal, Maharashtra, Karnataka, Tamil Nadu and Gujarat are the examples of the economic causes, while the influx from Myanmar and Pakistan to the states in the eastern and western parts of the country are the examples of political factors which encouraged in-migration into India.

The planning commission working group on population policy has estimated that the country's population will reach a figure of 1270 million in 2050. This would have serious consequences of food, energy, housing, employment etc. The programme for checking the rate of growth of population must be implemented properly. Unfortunately, wellplanned family planning programme is not successful here. It is possible only with full co-operation of both the central and state governments, volunteer organisations and the people at large.

REFERENCES

Agarwala, S.N. (1977), *India's Population Problems*, Tata McGraw-Hill, Delhi.

Bose, A. (1967), *Patterns of Population Change in India, 1951-61*, Allied Publishers,, Bombay.

Bose A. (1988), *From Population to People*, B.R. Publishing Corporation, New Delhi (2 Vols.).

Krishan, Gopal (1975), Some aspects of population growth in India: 1961-71, *Pacific View Point*, Vol. XVI, pp. 207-15.

Krishan, Gopal and Surya Kant (1985), Spurts in India's population growth during the twentieth century: A spatial view, *Population Geography*, Vol. VI (1 & 2) pp. 12-25.

Lahiri, Manosi (1981), Trends in urban growth in India, 1901-71, *Geographical Review of India*, Vol. 43(4).

Nag, P. (1984), India, in P. Nag (ed.), *Census Mapping Survey*, Concept Publishing Company, New Delhi for the IGU Commission on Population Geography, pp. 93-108.

Premi, M.K. (11982), The demographic situation in India, Paper of the East-West Population Institute, No. 8, Hawaii.

Roy, B.K. (1987), Demographic transition and growth of population in India, *National Association of Geographers of India Journal*, Vol. 33(1), pp. 17-28.

R.G.I. (1981), *General Population Tables, Series I, India*, Part II A (I)/A (II)/A (III), Census of India Publication.

14

COMPOSITION OF POPULATION

The composition of population, understood in terms of the ascribed and achieved attributes of the people, provides basic information about the social, economic and demographic structure of the population. The composition of population of India involves a vast field consisting of marital status, level of literacy and education, age, sex, religious composition, ethnic characteristics and language.

Ethnic Composition

India is the melting pot of various races and tribes. Almost all the major races of the world are present in the country and the tribal population is unbelievably varied and unique (Table 14.1).

The major races of India are Caucasoid, Mongoloid and Australoid. Negrite race is found in limited parts of the country like Bay of Bengal Islands, eastern and southern India. Caucasoid came to this country in 2000 B.C. Mongols are Asiatic in origin and they entered India both from the west and the east. At present they exist relatively in original form in eastern India, particularly in West Bengal, Assam and other northeastern states.

Australoid races in India are observed mainly in the south, namely in Tamil Nadu, Andhra Pradesh, Madhya Pradesh, Chota Nagpur and parts of the Ganga valley. The debatable Dravidian race who are sometimes linked to the Australoid race and sometimes to individual type are found in Malayalam and Tamil states. Tribal life in India is immensely varied. The major groups are (a) Dravidian tribe, (b) Mongoloid tribe, and (c) Turko-Iranian tribe. The Dravidian tribes are identified with the Mundas and the Santals of Chota Nagpur, and the Khond Mals of Orissa. These people are generally divided according to totems which bearing the names of a plant or animal common in the locality. These totems are considered to be the origin of that group. The Mongoloid tribe is found in the eastern part of India like the Nagas in the Eastern Hills and Chakmas of the eastern Himalaya. They believe in exogamy and practice totemism.

TABLE 14.1: Population of schedule castes and scheduled tribes, 1981

(*million*)

State/ Union Territory	Total population	Scheduled castes population	Percentage of S.C. population to total population	Scheduled tribes population	Percentage of S.Tribes population to total population
Andhra Pradesh	53.55	7.96	14.87	3.18	5.93
Bihar	69.92	10.41	14.51	5.81	8.31
Gujarat	34.09	2.44	7.15	4.85	14.22
Haryana	12.92	2.46	19.07	-	-
Himachal Pradesh	4.28	1.05	24.62	0.20	4.61
Jammu & Kashmir	5.99	0.50	8.31	-	-
Karnataka	37.14	5.60	15.07	1.83≥	4.91
Kerala	25.45	2.55	10.02	0.26	1.03
Madhya Pradesh	52.18	7.36	14.10	11.99	22.97
Maharashtra	62.78	4.48	7.14	5.77	9.19
Manipur	1.42	0.02	1.25	0.39	27.30
Meghalaya	1.34	0.01	0.41	1.08	80.58
Nagaland	0.78	-	-	0.65	83.99
Orissa	26.37	3.87	14.66	5.92	22.43
Punjab	16.79	4.51	26.87	-	-
Rajasthan	34.26	5.84	17.04	4.18	12.21
Sikkim	0.32	0.02	5.78	0.07	23.27
Tamil Nadu	48.41	8.88	18.35	0.52	1.07
Tripura	2.05	0.31	15.12	0.58	28.44
Uttar Pradesh	110.86	23.45	21.16	0.23	0.21
West Bengal	54.58	12.00	21.99	3.07	5.63
A. & N. Islands	0.19	-	-	0.02	11.85
Arunachal Pradesh	0.63	@	0.46	0.44	69.82
Chandigarh	0.45	0.06	14.09	-	-
D. & N. Haveli	0.10	@	1.97	0.08	78.82
Delhi	6.22	1.12	18.03	-	-
Goa, Daman & Diu	1.09	0.02	2.16	0.01	0.99
Lakshadweep	0.04	-	-	0.04	93.82
Mizoram	0.49	@	0.03	0.46	93.55
Pondicherry	0.60	0.10	15.99	-	-
All-India"*	665.29	104.75	15.75	51.63	7.76

* Excludes Assam where census could not be held owing to disturbed conditions prevailing there at the time of 1981 census.

" Excludes the population of area under unlawful occupation of Pakistan and China where census could not be held.

@ The figure does not come to even 2 decimal points when expressed in million

≥ This figure would appear to include high returns relating to certain communities with nomenclatures similar to those included in the list of scheduled tribes consequent on the removal of area restrictions.

Notes: 1. No castes were scheduled by the President of India for Nagaland, Andaman and Nicobar Islands and Lakshadweep.

2. 1981 figures of S.C. and S.T. are the enumerated figures based on S.C. and S.T. (Amendment) Act, 1976 and other relevant Acts and Orders.

3. No tribes were scheduled by the President of India for Haryana, Jammu & Kashmir, Punjab, Chandigarh, Delhi and Pondicherry.

Source: Census of India 1981, Series-I, India Part II B (i) & (iii) Primary Census Abstract for General, S.C. & S.T. Population.

The Turko-Iranian tribe may be divided into two groups : (i) Afghan groups, and (ii) Baloch-Brahui group. Afghan group, known as *Pathana* constitute of kindred group of agnates. The second group of Baloch-Brahui includes Brahuis, Balochs, Khetrans, Jats, Afghans, Kurds and Jagdals. There is no eponymous ancestor and the tribal faith is based primarily on common good or ill and not upon agnate kinship.

According to census enumeration, there are approximately 365 tribes in India. Jammu and Kashmir, Uttar Pradesh and Delhi have no schedule tribes whereas Andaman and Nicobar Islands, Arunachal Pradesh and Nagaland have no scheduled castes. Scheduled caste is present only in Hindu and Sikh religion. The share of scheduled castes and scheduled tribes to total population is 14.67 and 6.87 per cent respectively. The greater strength of schedule caste speaks of larger influence of the religions. Both these communities show predilection for rural areas and the cultural level reduces their number in areas. Sex ratio at national level of both scheduled castes and scheduled tribes is more favourable for woman.

The level of literacy for scheduled castes and tribes is 10.27 and 8.53 per cent respectively. Literacy for women is even lower. Occupational structure of these groups follow somewhat similar pattern as that of the general population. As rural residents, these people heavily depend upon agriculture. Most of the work force is cultivator or agricultural labourer while some are engaged in mining, forestry, livestock raising plantation, fishing etc.

Linguistic Composition

The multitude of races and tribes and their subgroups has given birth to a remarkable variety of languages. There are 1952 mother tongues of which 15 languages have been specified in the Eighth Schedule of the Constitution (Table 14.2). The fifteen major languages are of both Aryan and Dravidian origin. Aryan languages are found in use in northern India like Assamese, Bengali, Hindi, Punjabi, Kashmiri, Sanskrit and Urdu. Telugu, Kannada, Tamil and Malayalam are the Dravidian languages.

In the absence of written scripts, certain less common Dravidian languages are gradually dying away. However, the original language of India is said to precede Dravidian language. This is known as Kolerian language spoken by the various tribal people. Another language which is older than the Dravidian language is the Indo-Chinese language spoken by the people of the northeastern hills the Tibeto-Burmese population of the Himalayan areas.

TABLE 14.2: Language mainly spoken in households (by head of household)

Language	Population	Per cent
Assamese	70,525	0.01
Bengali	51,503,085	7.79
Gujarati	33,189,039	5.02
Hindi	264,189,057	39.94
Kannada	26,887,837	4.06
Kashmiri	3,174,684	0.48
Malayalam	25,952,966	3.92
Marathi	49,624,847	7.50
Oriya	22,881,053	3.46
Punjabi	18,588,400	2.81
Sanskrit	2,946	-
Sindhi	1,946,278	0.29
Tamil	44,730,384	6.76
Telugu	54,730,227	8.20
Urdu	35,323,282	5.34
Other languages	29,206,534	4.42

Source: 1981 Census.

The large number of enumerated languages indicated that the tribes of India maintained their individuality though isolated community lives in the historical past. Schedule tribe population speak their own mother tongue and 15-73 per cent of them practise bilingualism.

Maps 14.1 and 14.2 indicate the first and second languages spoken in each state according to the population. The second language gives some idea of the composition of population and migration trends.

Marital Status

Marital status of the Indian population is uneven for males and females in all age groups. In general the number of married women are greater than married men. The unmarried women are even fewer than the unmarried men. The social institutions in India rest upon collective economy and rarely on individual income. A divorced life for married couple means greater social insecurity for women under existing conditions. At present, very few married women are engaged in remunerative activities. The cases of widowhood is larger for the

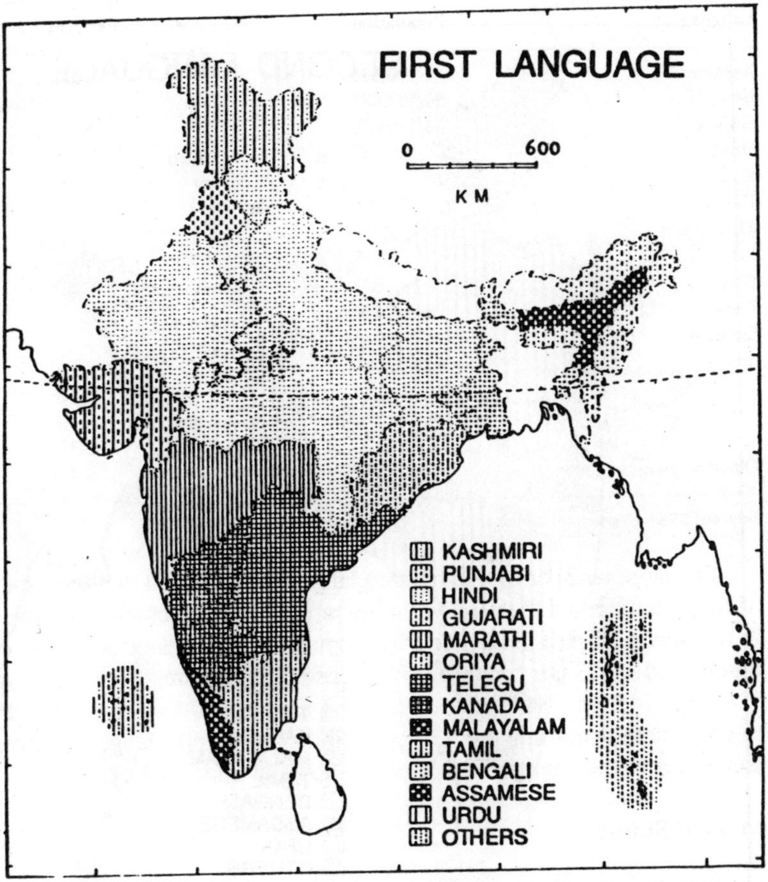

MAP 14.1: First Language

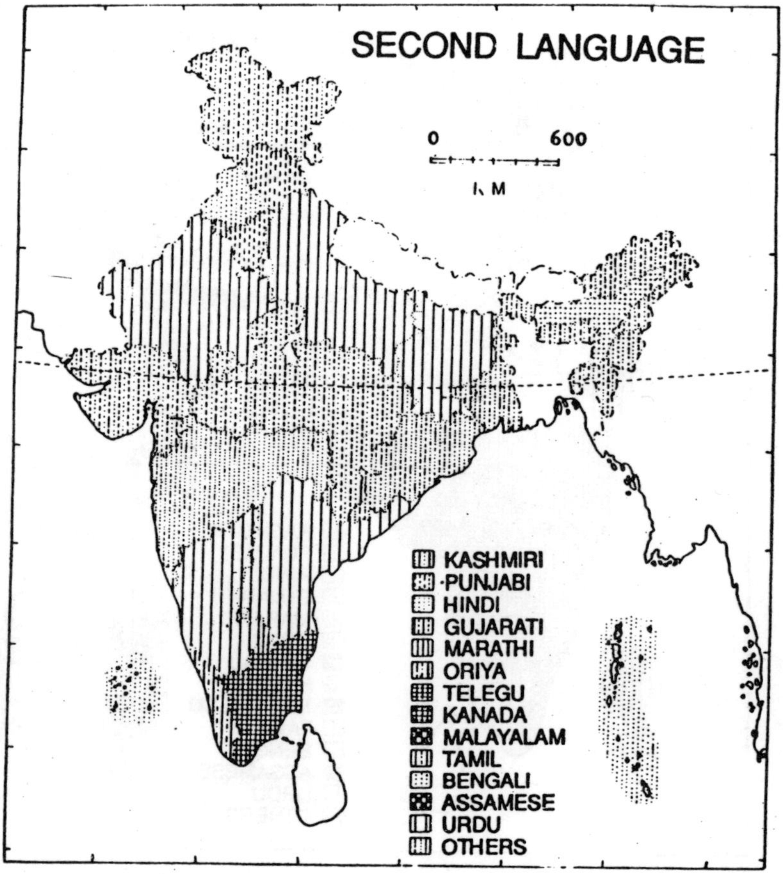

MAP 14.2: Second Language

women. But this is not due to the lesser expectancy of life for married men. In fact, widowers generally do not lead single life after bereavement. On the contrary widows in India do not marry for the rest of their life for social reasons. Though the modern law allows widow marriage, the social attitude does not encourage much such marriages.

Literacy

Literacy level is an important characteristic of population. There is a large increase in the literacy rate after independence. However, the overall literacy rate in the country was only 36 per cent in 1981 (Table 14.3). The result is more disappointing when female literacy is shown separately.

TABLE 14.3: Total population, literates and illiterates: 1901-81, India*

(in million)

Year	Population	Literates	Illiterates	Literacy rate
1	2	3	4	5
1901	235.1	12.6	222.5	5.3
1911	248.2	14.7	233.5	5.9
1921	246.7	17.7	229.0	7.2
1931	273.4	26.0	247.4	9.5
1941	312.0	50.2	261.8	16.1
1951	353.1	58.9	294.2	16.7
1961	428.1	102.6	325.5	24.0
1971	533.5	157.3	376.2	29.5
1981	665.3	241.0	424.3	36.2

* Excludes Assam.
Excludes population (297.853), literates and illiterates of that portion of Arunachal Pradesh where the all-India Census Schedule was not canvassed in 1961.

Note: Numbers of literates and illiterates for years between 1901 and 1941 were estimated by applying the literacy rate for undivided India to the population of that part of undivided India that remained in India after partition.

India's literacy rate has increased very much between 1901 and 1981. There is a great deal of variation in literacy rates among the states and union territories. Though, till now, the female literacy and rural literacy rates are comparatively low.

Kerala holds the distinction of highest literacy rate for both sexes. Gujarat, Maharashtra, Karnataka, Tamil Nadu and West Bengal were the other states showing literacy rates higher than the all India average. In all the states the growth rate of literates was considerably higher of female compared to that of males during the three decades starting from 1961.

The female and male literacy rates increased between 1961 and 1981 of every states and union territories. But when the different between male and female literacy rates is taken, a somewhat different picture emerges. The difference is more in states or union territories with low literacy rates and less in states or union territories with high literacy rates. Further, there is a considerable difference in literacy rates for urban and rural areas. The rural areas reclassified as urban tend to have higher literacy rates than the rural average but lower literacy rates than the urban average. The ranking of states and union territories by urban and rural literacy rates are quite similar (Table 14.4).

Age and Sex

In India, like other developing countries the data regarding age are not very reliable. Errors are due to misstatement of age. The population of young persons (below 15 years) increased from 39.3 per cent in 1921 to 41.9 per cent in 1981, for males and from 39 per cent to 42.2 per cent for females during the same period. The decrease in infant and child mortality seems to be the main cause for this increase in the young population. In all census years, the percentage of females in the old age groups, 60 years and over, was higher than that of males. There is an increase in the proposition in the young and old age groups between 1921 and 1971 in all states except Jammu and Kashmir, Kerala and Tamil Nadu. The highest proportion of old people in 1981 was found in Uttar Pradesh, followed by Punjab, Andhra Pradesh and Kerala. Most of the states showed a higher proportion of females than males in the old age group indicating lower female mortality in the advanced age. The dependency ratio for India in 1981 was 270 which is considerably high compared to that of a few earlier decades. Decreasing mortality unaccompanied by considerable decrease in fertility explains partly the increasing trend in the proportion of dependents (Table 14.5).

TABLE 14.4: Literacy rates and ranks by rural/urban residence for states and union territories, 1981.

India/state/ union territory	Literacy rate		Rank	
	Rural	Urban	Rural	Urban
1	2	3	4	5
INDIA	29.7	57.4		
States				
1. Andhra Pradesh	23.2	52.0	24	27
2. Bihar	22.5	52.2	26	26
3. Gujarat	36.2	60.3	15	15
4. Haryana	30.3	56.9	20	17
5. Himachal Pradesh	40.4	67.4	9	4
6. Jammu & Kashmir	21.6	45.6	27	30
7. Karnataka	31.0	56.7	19	18
8. Kerala	69.1	76.1	1	1
9. Madhya Pradesh	21.2	54.0	28	23
10. Maharashtra	38.2	63.9	13	10
11. Manipur	37.4	52.4	14	25
12. Meghalaya	27.5	64.1	22	9
13. Nagaland	38.6	64.2	10	8
14. Orissa	31.5	54.8	18	21
15. Punjab	35.2	55.6	16	19
16. Rajasthan	18.0	48.4	30	28
17. Sikkim	30.0	54.9	21	20
18. Tamil Nadu	38.6	63.4	11	11
19. Tripura	38.2	73.7	12	3
20. Uttar Pradesh	23.1	45.9	25	29
21. West Bengal	33.1	62.7	17	13
Union territories				
1. A & N Islands	46.6	65.5	7	6
2. Arunachal Pradesh	18.5	53.2	29	24
3. Chandigarh	44.7	66.2	8	5
4. Dadra & Nagar Haveli	24.7	54.2	23	22
5. Delhi	47.6	62.6	6	14
6. Goa, Daman and Diu	52.7	65.0	3	7
7. Lakshadweep	52.0	58.7	4	16
8. Mizoram	55.2	74.1	2	2
9. Pondicherry	48.0	63.0	5	12

* Exclude Assam.

TABLE 14.5: Age, sex composition, 1981

Age Group	Persons	Males	Females
0-4	12.55	12.28	12.85
5-4	14.08	14.03	14.13
10-14	12.92	13.16	12.65
15-19	9.64	9.89	9.37
20-24	8.62	8.43	8.82
25-29	7.62	7.49	7.77
30-34	6.37	6.28	6.47
35-39	5.84	5.79	5.90
40-44	5.14	5.24	5.03
45-49	4.39	4.47	4.31
50-54	3.82	4.02	3.61
55-59	2.47	2.47	2.46
60-64	2.73	2.73	2.73
65-69	1.43	1.39	1.47
70	2.33	2.28	2.38
Not stated	0.05	0.05	0.05
Total	100.00	100.00	100.00

Source: 1981 Census

Sex ratio (number of females per 1000 males) in India has declined from 955 in 1921 to 932 in 1981. The pattern of mortality, due to the higher risks of death for females during child birth, seems to be an important reasons for the abnormal sex ratio in India's population. The general trend of decline in sex ratio in the total population was also observed in the broad age groups 15-69 and 60 years and above. Andhra Pradesh, Bihar, Kerala, Madhya Pradesh, Karnataka, Orissa showed sex ratio above the all-India average in 1981. Among children below 15 years, the sex ratio in most of the states did not differ much throughout this century. Among the old people (60 years and above) there was a surplus females in 1981. Females, having overcome hazards of the maternity, live much longer than males in old age. During 1921-81 there was an increase in the sex ratio in the 15-59 age group in some states. This may be due to the selective male outmigration from these states.

Religious Composition

Religion is an important characteristic of population. Society in India is still largely organised on the basis of caste and religion which affect the social, economic and political structure of the society and thereby influence the course of population change. Social behaviour is much influenced by practices which are subject to religious sanctions. It is observed that the age of marriage, the practice of remarriage, the occupation of women, etc. vary according to the caste and religious affiliation of the individual.

Hindus accounted for the highest proportion (85%) of India's population in 1981 (Table 14.6). But their growth rate in different decades in this century is below the growth rate of the general population. In 1981, Jammu and Kashmir was the only state in which less than 50 per cent of the population was Hindus. The proportion of Muslims in the country increased in all the censuses except for the decade 1941-51. This decrease was due to the partition of the country. Jammu and Kashmir is the only state where Muslims were in a majority. In 1981, about half of the total Muslims population of India was distributed in three states : Bihar, Uttar Pradesh and West Bengal.

TABLE 14.6: Religious composition

Religion	Population	Per cent
Hindu	549,724,717	82.63
Muslim	75,571,514	11.36
Christians	16,174,498	2.43
Sikhs	13,078,146	1.96
Buddhists	4,719,900	0.71
Jains	3,192,572	0.48
Other religious	2,766,285	0.42
Not stated	60,217	0.01

Source: 1981 Census.

The Christians, third major religious community in India, accounted for 2.6 per cent of India's population in 1981. About two-thirds of the total Christians in India were living in the southern states i.e. Andhra Pradesh, Kerala, Karnataka and Tamil Nadu. Population increase among Christians in different states has followed no consistent pattern. The proportion of Sikhs in the total population of India has increased considerably in this century. Their growth rate also quite high like that of the Christians. About 86 per cent of all the Sikhs in India in 1981 reside in Punjab and Haryana. Buddhists and the Jains constituted less than one per cent of India's population in 1981. The enormous increase of Buddhists in 1961 is attributed to the

New-Buddhists movement, and migration of Tibetans due to Chinese invasion in their country. Jainism in India is mainly concentrated within two linguistic groups in this country. So they are seen mostly in Gujarat and Rajasthan. It is observed that some of the minor religious communities in different states of India are growing faster than the major religious groups (Appendix X).

India present a unique picture with so much variations in its population composition. Though the people of India lived in a harmony with their diversity for centuries but in present situation communal, ethnic and linguistic divisions are creating tension in some parts of the country. In some cases this is due to the political and social awareness while in others it is due to the unemployment and political unstability in the neighbouring countries. The political unstability in the neighbouring countries is increasing illegal migration across in international border and creating tension in this country.

REFERENCES

Chandra, R.C. (1989), *Spatial Dimensions of Schedule Castes in India*, Chandigarh.

Mukerji, A.B. & Mehta, Swamjit (1975), Female participation in agricultural labour in India, pattern and association, *Tijdschrift voor Economische in Social Geografic*, Vol. LXVI, pp. 103-7.

Narain, Vir (1985), Marginal worker in India, *Demography India*, Vol XIV (2), pp. 227-35.

Nayak, D.K. & Aijazuddin Ahmad (1985), Female participation in economic activity : A geographical spectrum with special reference to rural areas in India, *Indian Geographical Journal*, Vol. LIX(2).

Patnaik, Indu (1985), Changes in the distributional pattern of workers in manufacturing in India : 1901-71, *Population Geography*, Vol VII (1 & 2), pp. 18-26.

Raza, Moonis, Aijazuddin Ahmad & S.C. Nuna (1985), Spatial patterns of tribal literacy in India, *The Indian Geographical Journal*, Vol LX (2), pp. 103-16.

Raza, Moonis & Sudesh Nangia (1986), *Atlas of the Child in India*, Concept Publishing Company, New Delhi.

RGI (1988), *A Handbook of Population Statistics*, Census of India 1981 Publication, New Delhi.

Singh, K.S. (ed.), (1972), *Tribal Situation in India*, Indian Institute of Advanced Study, Shimla.

15

MIGRATION PATTERN

According to the Indian Census, the migration is determined by the concept of place of birth or residence. The census does not account for the intermediate movement of persons between various places. But if a person was born at a place other than the place of enumeration then he will be figured as a migrant. This concept of migration may not provide a true picture and may lead to a different meaning of this word. On the other hand, according to the Indian tradition, most of the first child of a family are born in the homes of grandparents. Only for this custom a person cannot be enumerated as a migrant in his native places. However, migration or migrants have a wider meaning.

Some tribal groups of central Himalaya and shifting cultivators of eastern Hills and central plateau practise regular movement. They can be termed migrant in the real sense.

International Migration

International migrants in India are few, compared to total population. Most of the international migrants are from Asian countries, followed by Europeans, Africans etc. The neighbouring countries like Pakistan, Nepal, Bangladesh, Afghanistan, the USSR have large number of people in India (Table 15.1).

An examination of the list of international migrants reveals that many countries have no significant historical or cultural relations with India. These migrant populations exclude persons employed in the embassies and missions. So, some more countries may have connections with this country besides the enlisted record.

Sex-composition of the non-Indian nationals to this country indicates business-bound movement. In almost all cases males out number the females. Everywhere in the world, business-bound movement carries more males, in some countries the picture is different, e.g., Ireland, Mozambique, Canada etc. Irish females in this country come on various missionary activities and also through marriages. In case of Mozambique this is due to slight majority of females and their family composition. More Afghan and Nepalese males are due to the social structure which force them to lead a hard and isolated life into order to earn money for their family at home.

TABLE 15.1: International migrants, 1981 (percentage to total international migrants)

Place of last residence	Total	Male	Female
Countries in Asia beyond India	98.8	47.51	52.49
Europe	0.28	29.25	70.75
Africa	0.70	22.18	77.82
Americas	0.01	15.85	84.15
Oceania	0.02	21.07	78.93

Source : A Handbook of Population Statistics.

Distribution of non-Indian nationals varies within different states. Assam and Uttar Pradesh receive highest numbers of foreign nationals, much higher than that of West Bengal, Maharashtra, Delhi and Tamil Nadu, the states which have metropolitan cities. An analysis shows that from the neighbouring countries like Bangladesh and Nepal foreigners inflict to these territories which has a common cultural background. In fact, Assam receives more people from Bangladesh than the Nepalese, whereas Uttar Pradesh has more Nepalese. This pattern clearly reflects the importance of distance factors in the movement of population. Due to this factor large number of Nepalese are seen in Uttar Pradesh, Bihar, Punjab, Himachal Pradesh, Arunachal Pradesh, Maharashtra and Delhi. Obviously this international migration is due to better economic opportunities, which exist in these states of India.

Countries other than those adjacent ones however, send their people to urban, industrial, and plantation areas, like those of West Bengal, Maharashtra and Delhi. The countries which had political relations, along with economic, cultural and diplomatic relations, have more people in Maharashtra, West Bengal and Delhi. Other states like Madhya Pradesh, receives the largest number of the USSR nationals for its steel plants built with Soviet collaboration, and similarly Orissa had a sizeable West German nationals. Concentration of Arab nationals in the western states like Maharashtra, Gujarat, Goa, Daman and Diu and Kerala is mainly due to the proximity.

Inter-state and Intra-state Migration

The state level migration or inter-state migration is not very high in India. Almost everywhere the pattern of inter-state migration is high in urban areas. Which naturally form the alluring pools for economic and social opportunities (Table 15.2). Madhya Pradesh, Maharashtra, Orissa, West Bengal and Delhi have a higher number of migrants in their urban areas. This is mainly due to the higher rate of urbanisation, industrialisation and mining activities. A noticeable feature among the urban migrants is the deficiency of females. The urban centres of Laccadive, Minicoy and Amindivi Islands, Dadra, Nagar Haveli and Arunachal Pradesh have nearly 100 per cent outside population. It shows local people refrain from living in urban centres of these areas. In case of inter-state migration, the higher migration to urban areas depicts the long distance movements are generally urban oriented. Cities and towns everywhere bear certain similar cultural environment, and this similarity of urban culture presents a congenial atmosphere for the mobile population to take advantage of the opportunities available there.

Inter-state migration concerning rural areas is very low all over the country. In case of inter-state migration in rural areas the ratio of female is high in some places. It means that in this case the whole family have moved. In India the intra-state migration is higher in rural areas than the urban one. Though it is a common experience that urban magnet is more powerful for migrants than rural spheres, yet short distance movements involve more rural population representing cultivators, artisans and other persons engaged in other primary activities. Intra-state migrants to urban areas are smaller owing to the fact that the degree of urbanisation in India is low (Table 15.3).

Sex composition in both intra-state and inter-state migrations is characterised by higher number of females in rural areas and lesser number in urban areas. Movement to rural areas generally involves the whole family as work is low paying, people cannot afford keeping their families in native places and support them by remittances. On the other hand, when a job or work or business in urban areas pays well, people can leave their families in native places for economy and comfort since the cost of living is high and accommodation is a great problem in cities and towns (Table 15.4). But, the above situation is not always true. Families do join bread earners in urban areas when the income becomes reasonable. However, incomplete and nuclear families are becomings common in urban centres. Bigger the centre, larger the number of such families are to be found there (Maps 15.1 and 15.2).

Based on the rural or urban nature of the place of birth and the place of enumeration, internal migrants can be classified into four migration streams — rural to rural, rural to urban, urban to rural, and urban to urban (Table 15.5). Rural to rural migration formed the most dominant stream at the national level but its importance for both males and females declined over time. While a very large proportion of female intra-district rural to rural migration is due to marriage. Male migration is from areas of low agricultural productivity to areas of higher agricultural productivity with better wages, or to areas which have construction or other similar projects within the district.

TABLE 15.2: Migrants by place of birth — 1981 (in percentage of total migrants)

Birth Place	Status of Place of enumeration	Male	Female
Total	Rural	52.18	78.08
	Urban	47.82	21.92
A. Migrants born	Rural	58.96	82.14
With the state	Urban	41.04	17.86
(i) Born elsewhere	Rural	70.81	88.11
in the dist of enu.	Urban	29.19	11.89
(ii) Born in other	Rural	40.26	65.45
distt. of state	Urban	59.74	34.55
B. Born in other	Rural	26.89	45.65
states of India	Urban	73.11	54.35
C. Born in Countries	Rural	47.79	49.49
in Asia beyond India	Urban	52.21	50.51
D. Born in countries	Rural	28.08	23.39
in Europe	Urban	71.92	76.61
E. Born in countries	Rural	28.08	23.39
in Africa	Urban	79.40	79.22
F. Born in countries	Rural	16.30	16.46
in America	Urban	83.70	83.54
G. Born in countries	Rural	17.77	20.75
in Oceania	Urban	82.23	79.25
H. Unclassifiable	Rural	30.88	39.44
	Urban	69.12	60.54

Source: Handbook of Population Statistics.

TABLE 15.3: **Internal migrants by place of residence and duration of residence in place of enumeration**

(in per cent)

Place of last residence	Total		1 yr.		1—4 yrs.		5—9 yrs.		10—19 yrs.		20 yrs.		Period not stated	
	M	F	M	F	M	F	M	F	M	F	M	F	M	F
A. Elsewhere India	59,235 306	142,371 755	8.57	3.71	26.16	17.01	16.65	14.76	20.96	23.61	20.91	37.25	5.75	3.66
I. Within state	47,988 451	130,079 238	8.54	3.51	26.68	16.64	16.70	14.58	20.38	23.56	20.47	38.09	7.23	3.62
1. Elsewhere in distt.	30,296 162	96,939 721	8.16	3.11	26.29	15.75	16.42	14.20	19.74	23.58	20.95	39.71	8.44	3.65
2. In other distt.	17,692 289	33,139 516	9.19	4.66	27.34	19.27	17.19	15.68	21.48	23.49	19.64	33.37	5.16	3.54
II. Other states in India	11,246 865	12,292 518	8.71	5.94	23.98	20.92	16.41	16.62	23.43	24.17	22.80	28.43	4.67	3.92

Source: A Handbook of Population of Statistics.

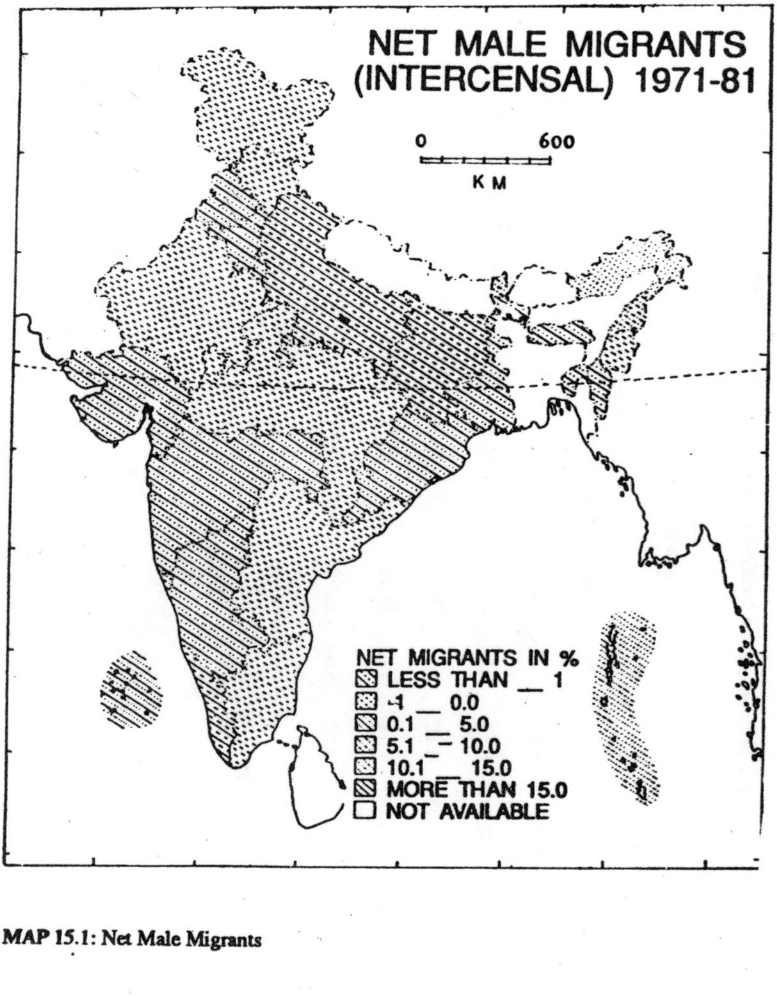

MAP 15.1: Net Male Migrants

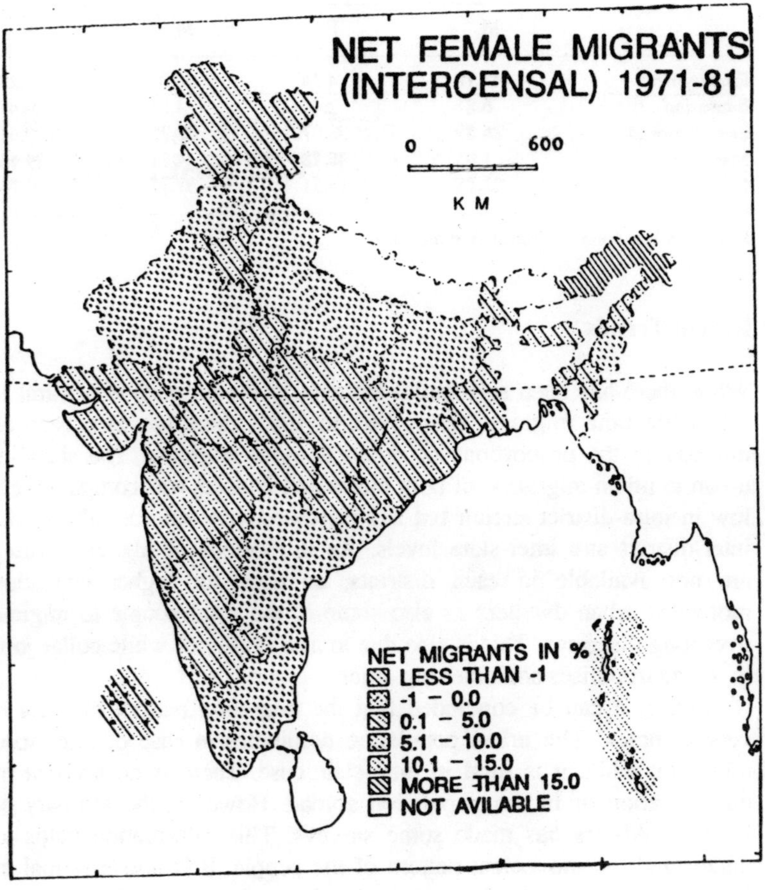

MAP 15.2: Net Female Migrants

TABLE 15.4: Reasons for migration

(in percentage)

Reason/Place	Urban		Rural	
	M	F	M	F
Employment	43.10	4.24	20.07	1.28
Education	6.81	2.40	4.06	0.47
Family moved	26.89	32.07	33.23	9.61
Marriage	1.05	46.78	4.87	79.46
Others	22.15	14.51	37.77	9.18

Source: A Handbook of Population Statistics.

Recent Trends

While there has been substantial increase in the proportion of rural to urban life time migrants from 1961 to 1981, there has also been an increase in the proportions of urban to rural migrants. The share of urban to urban migration of both males and females was comparatively low in intra-district stream but has increased very substantially in the inter-district and inter-state levels. As institutions of higher learning are not available in each districts, an urge for higher education motivates urban dwellers as also some of the rural people to migrate over long distances. This is also due to availability of white collar jobs in the metropolises and other big cities.

Hence, it can be concluded that the migration pattern in India is very complex. The urban population dominates in case of inter-state migration while it is rural in intra-state case. There is no account of out-migration of Indian population abroad. However, the Ministry of External Affairs has made some surveys. This information helps to understand the movement pattern of the people. It is also essential to have information according to the level of education and occupation and also income of the migrants. The socio-economic level of migrants is an important factor which controls the process and cause of migration.

TABLE 15.5: **Distribution of migrants**

(Percentage)

Type migration stream	1961		1871		1981	
	M	F	M	F	M	F
A. Intra-district (T)	54.45	73.78	53.58	71.59	48.91	67.00
Rural to rural	40.15	65.48	38.42	62.01	31.26	56.02
Rural to urban	9.03	4.32	9.42	5.19	10.65	5.82
Urban to urban	2.97	1.65	2.47	1.47	3.61	2.13
Urban to rural	2.30	1.83	3.27	2.92	3.36	2.96
B. Inter-district (T)	26.78	19.00	27.24	20.20	30.98	23.97
Rural to rural	11.28	12.44	10.14	12.23	10.05	13.80
Rural to urban	8.80	3.15	9.02	3.42	10.79	4.27
Urban to urban	5.22	2.45	6.00	3.04	7.68	3.99
Urban to rural	1.48	0.96	2.08	1.51	2.40	1.86
C. Inter-state (T)	18.77	7.22	19.18	8.22	20.11	9.04
Rural to rural	5.31	3.88	4.97	3.51	4.17	3.41
Rural to urban	7.83	1.76	7.55	1.91	8.53	2.36
Urban to urban	4.87	1.72	5.48	2.17	6.05	2.52
Urban to rural	0.76	0.36	1.18	0.63	1.21	0.70
D. All Streams (T)						
Rural to rural	56.74	81.30	53.53		[illegible]8	73.23
Rural to urban	25.66	9.73	25.99	[illegible]	[illegible].97	12.45
Urban to urban	13.06	15.82	13.95	6.69	17.34	8.64
Urban to rural	4.54	3.15	6.53	5.05	6.97	5.53

Source: Premi (1989)

REFERENCES

Gosal, G.S. and Gopal Krishana (194), Patterns of internal migration in India, in L.A. Kosinski and R.M. Prothero (eds.), *People on the Move*, Methuen & Co. Ltd., London, pp. 193-206.

Mitra, Ashok, Shekhar Mukherjee, & R. Bose (1980), *Indian Cities: Their Industrial Structure, Immigration and Capital Investment*, Abhinav Publishers, New Delhi.

Premi, M.K.. (1989), Pattern of internal migration in India: Some new dimensions, in S.N. Singh, M.K. Premi, P.S. Bhatia and A Bose (eds.), *Population in Transition in India*, B.R. Publishing Corporation, Delhi, pp. 277-92.

Raju, B.R.K., (1989), Development Migration : *A Processual Analysis of Inter-state Rural-Rural Migration*, Concept Publishing Company, New Delhi.

RGI (1986), *Proceedings of Workshop on Migration and* Urbanisation, (March 10-28, 1986), New Delhi.

Skeldon, R. (1985), Migration in South Asia: An overview in L.A. Kosinski & K.M. Elahi (eds.), *Population Re-distribution and Development in South Asia*, Reidal, Boston, pp. 37-63.

16

RURAL SETTLEMENTS

India is predominantly an agricultural and rural country. An overwhelming majority of her people (76.7%) live in the countryside. From time immemorial the village has been a basic and important unit in the organisation of Indian social set up.

Today it is common to speak of the Indian village. It is true that many problems are common to entire countryside, but the village as a social and cultural unit possesses a basically uniform organisation and structure all over the country. This would obviously be an over simplified description of a complex situation. To understand their social structures as well as their problems it is necessary to classify them into clearly distinguishable types.

Each cultural area of India has its own nomenclature to classify the villages within its territory in respect of their size and population. In the Hindi speaking parts of north and central India it is usual to distinguish between *Kheda* (a hamlet), *gaon* (a small village) and *Kasba* (a large village). In the southern part of the country, a similar distinction is made between *gumpu, majra* and *grama*. The first of these denotes a small collection of huts, the second is a larger hamlet and the third is the entire village. The differences between these rural settlements are confined not only to their size and population. They differ greatly in their internal organisation, and economic self sufficiency.

The ethnic and linguistic composition of a village as well as its caste structure determines its character. The structure, ethos and problems of a wholly tribal village will be different from those of a village with a mixed population. A tribal village is often unitary with respect to its structure. The traditional tribal authority is supreme in functional, ritual and cultural matters. On the other hand, in the case of a village which has a mixed population having different culture, language and religion, the nature of its internal organisation and authority is bound to be different. Location of a village in relation to administrative, industrial centres, and transport nodes also have some bearings on the nature of the organisation and modifies the setting and social fabric of a village.

The rural settlements including layout and house type are governed by certain traditions — regional and local. Hence, in view of these factors, it is difficult to regard any one village as typical or representative of rural India as a whole. It is essential to study village communities from different parts of the country to get a true picture of rural India. Though attempts have been made to study the role of caste specific hamlets, traditional pollution distance, *jajmani* systems, dominant social group, territorial character and patterns of transformation of the Indian villages.

Size of Villages

The distribution of the villages by population (Table 16.1) shows the predominance (48.6%) of small size villages (less than 500 persons). The village size between 200 to 499 persons is predominant. The percentage of villages decline in successive groups. This aggregate pattern of course varies from the state to state, mainly due to the diversified physiography, climate and cultural set up. The percentage of small size villages (less than 200 persons) is higher in hilly region, for example Himachal Pradesh (62.73), Manipur (46.09) and Meghalaya (65.29). The hilly mountainous terrain is not suitable for large settlements. On the other hand, states with more or less plain topography have large number of medium size villages.

The physiography and environment play an important role in the locational arrangement of villages with respect to one another. In the hilly and unsuitable tracts e.g. in Jammu and Kashmir, Sikkim, Arunachal Pradesh, Mizoram the spacing is very high and it some time reaches up to 5 km (Map 16.1); while in the Ganga Plains, i.e. Uttar Pradesh, is low (less than 2 km). It can be said that the low spacing with high density is the character of the plains, while high spacing with low density are the character of heterogeneous environment (Table 16.1).

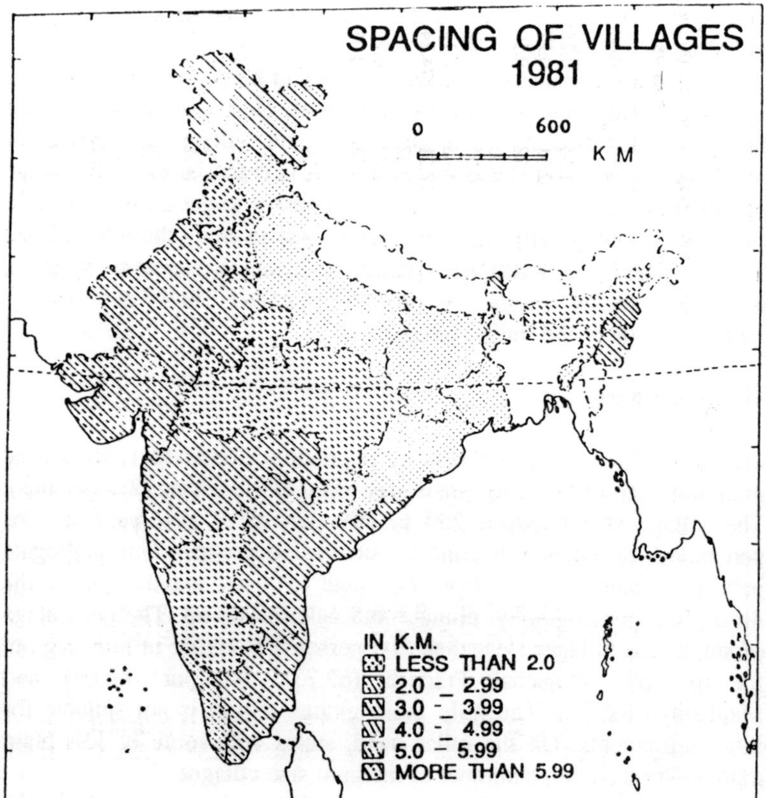

MAP 16.1: Spacing of Villages

TABLE 16.1: Villages, their size, spacing and percentage of villages with population (1981)

States/Union Territories	Total No. of inhabited villages	Population size							Spacing km
		Less than 200	200-499	500-999	1000-1999	2000-9999	5000-9999	10000 and above	
Andhra Pradesh	27,379	17.50	14.87	14.04	23.60	20.48	3.86	0.60	3.42
Assam	21,995	—	—	—	—	—	—	—	2.03
Bihar	67,546	21.89	28.19	23.38	16.05	8.65	1.56	0.28	1.72
Gujarat	18,114	7.65	19.17	23.36	28.20	15.04	2.31	0.27	3.52
Haryana	6,745	6.89	15.33	26.33	27.92	19.96	3.28	0.30	2.76
Himachal Pradesh	16,807	62.73	27.13	7.40	2.16	0.55	0.02	—	1.96
Jammu & Kashmir	6,477	19.68	31.29	26.47	16.15	6.06	0.31	0.02	6.28
Karnataka	27,028	15.21	26.55	27.16	19.38	10.07	1.48	0.13	2.87
Kerala	1,219	0.25	0.16	0.16	1.07	5.90	18.21	74.24	5.95
Madhya Pradesh	71,352	22.15	36.36	27.02	11.38	2.87	0.22	—	2.69
Maharashtra	39,354	11.49	23.50	31.33	22.75	9.16	1.47	0.29	3.15
Manipur	2,035	46.09	27.32	12.97	8.55	4.42	0.58	0.05	3.64
Meghalaya	4,902	65.25	26.31	6.20	1.73	0.49	—	—	2.38
Nagaland	112	26.70	34.44	21.86	12.86	4.22	0.09	—	4.47
Orissa	46,553	39.21	32.13	9.79	2.65	0.10	—	—	1.96
Punjab	12,342	12.72	23.19	30.07	22.84	10.17	0.97	—	2.19
Rajasthan	34,968	22.48	29.81	24.41	15.63	6.78	0.92	0.60	3.45
Sikkim	440	13.18	40.23	31.13	13.63	1.82	—	—	4.50
Tamil Nadu	15,831	5.01	10.94	20.03	28.86	27.67	6.32	1.15	3.09
Tripura	856	3.97	10.63	16.35	29.55	32.00	6.54	0.93	1.60
Uttar Pradesh	112,566	19.89	27.38	26.67	18.10	7.24	0.67	0.05	1.74
West Bengal	38,024	16.22	25.65	24.71	19.82	11.53	1.84	0.22	1.64
Union Territories	5,987	57.39	19.72	9.62	7.38	4.76	1.11	0.02	3.85
Total/All India	557,137	21.55	27.05	24.39	16.96	8.41	1.29	0.33	2.57

Source: Indian Agriculture in Brief. Ministry of Agriculture, New Delhi, 21 Ed, 1986, pp. 200-01.

Types of Rural Settlement

There are various factors which affect settlement types in rural areas. The factors may be both attractive or repulsive in nature. The older settlements were influenced by various physical factors like topography, fertile soil, side of the slopes favourable climate etc., but now economic and cultural factors influence the settlement type greatly. Economic and cultural factors are mainly transport routes, smaller business centres, places of tourist interests and religious or cultural divisions.

Settlements generally develop with reference to waterbodies or constant source of water. Hence in areas where people have to depend on wells or tanks like semi-desert and desert regions in west and peninsular regions in central part, there the settlements are generally compact. However, in areas where water table is high and wells can be dug easily or where the drainage texture is dense, settlements can be built practically all over the region. In such cases, hamlets, semi-compact settlements are more common. Such are as are found in the Great Plains.

Compact settlements in India are also due to the agricultural activities. The type of agriculture which involves large labour force requires both farmers and the agricultural labours to reside in the same nucleated settlements. Due to the lack of transport facilities till very recently, the economic self-sufficiency was a major factor leading to the formation of compact settlements.

In areas where the physical factors do not prevent dispersal and security is not a problem, small hamlets rather than large clusters came into existence. Sometimes the emergence of a market site around a railway junction or a railway station, a little away from the main settlement attracts a few dwelling units as well. This in turn creates a separate hamlet. In most cases, during the course of years, this hamlet outgrows and becomes an independent settlement with viable economic base.

Rural Settlement Pattern

Villages in India may be considered as a natural outcome of physical and cultural setting. Although they do not possess well-defined shapes and a distinct internal plan, there is some pattern, both internal and external, which can clearly be related to the nature of their site and arrangement. Village cart tracks, roads, mosques and temples are intimately related to the structure of the village. The state of insecurity in the past and the present social composition of the villages are significant factors in the development of rural settlement pattern. Nevertheless, in a large number of villages, lanes are so crooked and dwellings arranged in so irregular a fashion that no pattern is found *per se*. The most common shape of the village is roughly rectangular. One of the main reasons for this pattern is the original rectangular shape of cultivated fields. The system of land measure based on some sort of square unit *(bigha)* has resulted in a similar field pattern.

In a large rectangular village lanes intersect roughly at right angles and combined with the rectangular subblocks of the village presents a checker board pattern. Generally the village is made up of several wards inhabited by distinct castes. In the building of houses one of the primary consideration is accessibility to the nearest well.

Another frequent form of pattern is of the elongated village where one axis of the village is markedly longer than the other. Natural or cultural forces on the site are restricting its extension on one axis. This pattern is found along the higher ground level in inundated areas, narrow strip between two streams and at the edge of an alluvial terrace. Among cultural features the most important is a road. If the road is an important link with the surrounding villages or towns, then the elongation is obvious. The extension of a village along the road is often remarkably encouraged if the settlement is a market centre. Other common pattern of Indian villages are fan pattern, circular village, polygonal village, oval village, horse-shoe pattern (plateau regions) double nucleations, irregular cluster etc.

Pattern which depends on the arrangement of streets and houses cannot be studied in the case of dispersed settlements where dwellings are scattered and streets are non-existing. There is, however, some arrangement noticeable in the placing of dwellings and on this basis two types are found viz., string or linear (canal-irrigation dry area) pattern and contour pattern (in the hilly region).

The distribution of population, setting and types of rural settlements, village patterns are to a great extent related to the natural and cultural features of the area (Fig. 16.2).

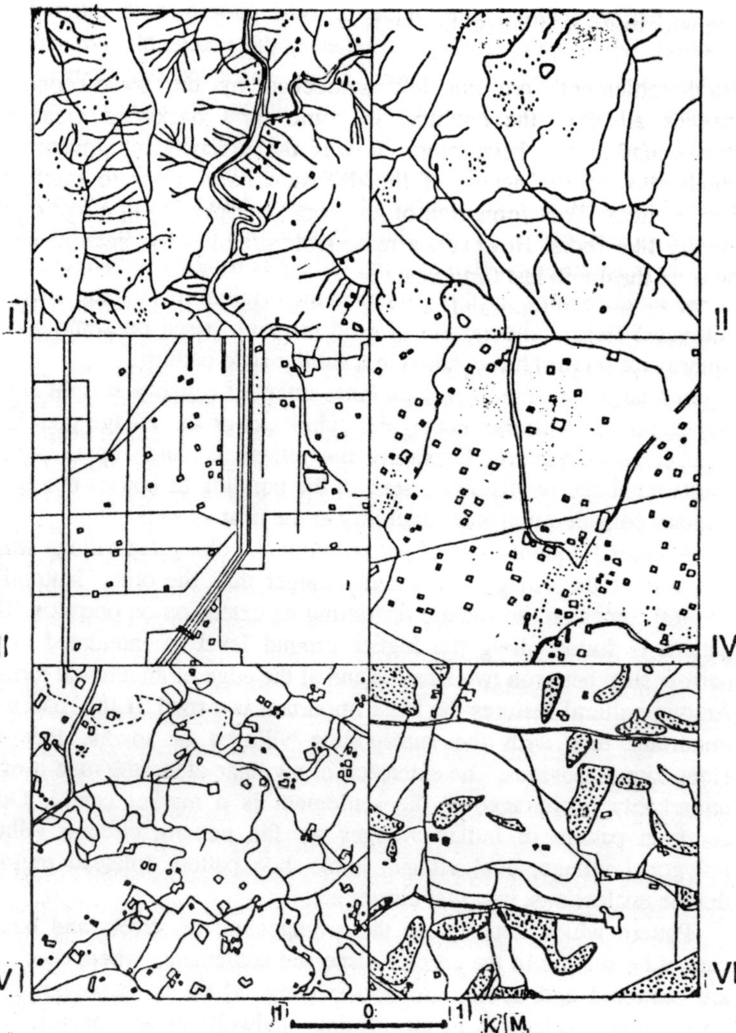

I. Dispersed settlement in Himachal Pradesh
II. Semi-compact Settlements in Madhya Pradesh
III. Canal Settlements in Rajasthan
IV. Settlements in Bengal
V. Settlements in Ganga Plain
VI. Tank-Settlements in Madurai

FIG. 16.2: Rural Settlement Patterns

Transformation of Rural Settlements

Rural settlements are not left untouched by the socio-economic changes all over the country. In some rural backward areas the transformation has been more. It was the independent government which allowed the agents of the diffusion innovations to reach the interior areas. Transformation of any sort is purplexing in a traditional country like India. However, a reasonable progress has been made in the rural theatre in last forty years or so.

There are 576 thousand villages in the country, out of which about three-fourths are electrified. Obviously, these villages are relatively better located. Further, in smaller states and union territories, all the villages were electrified in early 1988. These are Haryana, Punjab, Chandigarh, Dadra & Nagar Haveli, Delhi, Lakshadweep and Pondicherry. There are some states which have over 90 per cent electrified villages. States like Bihar, Orissa and West Bengal have less than 60 per cent electrified villages (Table 16.2). In such villages it is not necessary that all the households are having electricity. A large proportion of electricity is consumed for irrigating the fields. In 1984-85, about 18.7 per cent of the total consumption of electricity was in agricultural sector. The anticipated number of electrical pumpsets by the end of the Sixth Plan (1985) was 5.73 million, i.e., on an average of about 10 sets per village. The target in the Seventh Plan (1985-90) was to have an additional 2.35 million such sets. In addition, there are diesel pumpsets as well. Up to the end of the Sixth Plan, the anticipation number of sets was 3.55 million. However, the target by the end of the Seventh Plan was one million sets.

Another modern amenity is the rural roads. Earlier some forms of tracks did exist which was not always suitable for transformation. Number of villages connected by rural roads by the end of the Sixth Plan was 202 thousand, i.e., less than half of the villages. The Seventh Plan target was for only 24 thousand villages, mostly having a population size more than 1,500 person (Table 16.3). There were high percentage of villages which were away from metalled road (43.7% - 8 km or more), railway station (55.8% - more than 15 km), and post office (65% - 8 km or more) (Table 16.4).

Electricity and roads have brought certain changes in villages. The life style is transforming reasonably faster. Televisions, videos, tractors, cars, motor-cycles, scooters, electric appliances, fertilisers and the like are becoming common. However, such transformation is

TABLE 16.2: Village electrification

States/Union Territories	Villages	Percentage of villages electrified as on	
		31-3-85	31-1-88
Andhra Pradesh	27,221	83.96	91.6
Arunachal Pradesh	2,973	27.82	—
Assam	21,995	53.67	75.6
Bihar	67,566	49.93	58.4
Goa	—	—	—
Gujarat	18,275	88.29	99.5
Haryana	6,731	100.00	100.00
Himachal Pradesh	16,916	86.39	99.5
Jammu & Kashmir	6,503	87.73	92.3
Karnataka	26,824	84.50	97.5
Kerala	1,268	100.00	100.00
Madhya Pradesh	70,883	57.10	70.1
Maharashtra	35,778	92.84	95.1
Manipur	1,949	30.89	—
Meghalaya	4,583	27.54	—
Mizoram	229	95.41	—
Nagaland	960	70.83	—
Orissa	46,992	50.57	58.3
Punjab	12,188	100.00	100.00
Rajasthan	33,305	59.38	64.6
Sikkim	405	46.67	—
Tamil Nadu	15,735	99.78	99.4
Tripura	4,767	39.45	—
Uttar Pradesh	1,12,561	56.04	65.3
West Bengal	38,074	50.43	59.8
Andaman & Nicobar Island	390	62.05	—
Chandigarh	26	100.00	—
Dadra & Nagar Haveli	72	100.00	—
Delhi	243	100.00	—
Goa, Daman & Diu	409	92.91	—
Lakshadweep	10	100.00	—
Pondicherry	333	100.00	—
Total	5,76,126	64.11	73.6

Source: 1. Indian Agriculture in Brief
2. Statistical Outline, TCS.

different from Gandhiji's dream of Indian villages. Material culture, luxuries and other unproductive investments are the signs of negative aspects of the urbanism which is creeping in the rural scenario.

Bharlai is an example of a traditional Indian village in Ganga plain now fast transforming into a fringe of the Varanasi City (Fig. 16.3). Landuse and social set up have gone rapid changes in the recent past.

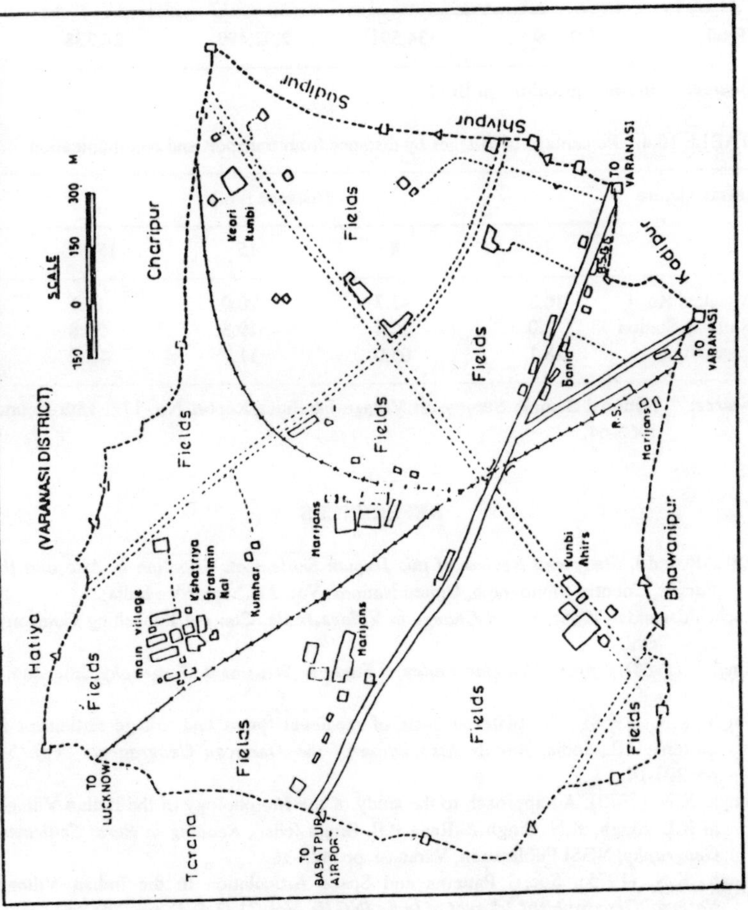

FIG. 16.3: Bharlai Villages

TABLE 16.3: Rural roads programme — accessibility of villages with roads

Population category	Total No. of villages	No. of villages covered during Sixth Plan	No. of villages connected as on 31-3-85	Target for 7th Plan as per material norms
1500+	69,408	10,924	48,921	20,487
1000-1500	56,609	6,839	28,967	3,851
Less than 1000	4,66,075	16,828	1,24,611	—
Total	5,92,092	34,591	2,02,499	24,338

Source: Indian Agriculture in Brief.

TABLE 16.4: Percentage of villages by distance from transport and communication

Infrastructure	Distance (km)			
	1	8	15	15+
Metalled Road	16.3	43.7	20.0	19.1
Railway Station	1.0	20.4	19.8	55.8
Post Office	18.1	65.0	11.7	3.9

Source: National Sample Survey on Village Facilities Report No. 172, 18th Round, 1963-64.

REFERENCES

ESCAP (u.d.), *Study and Review of this Human Settlements Situation in Asia and the Pacific*, Country Monograph, United Nations, Vol. 2 (Chapter V : India).

Sachchidananda (1988), *Social Change in Village India*, Concept Publishing Company, New Delhi.

Sing, J. (1985), *Upper Damodar Valley : Study in Settlement Geography*, Inter-India, New Delhi.

Singh, K.N. (1968), The twitorial basis of medieval towns and village settlement in Eastern U.P., India, *Annals Association of the American Geographers*, Vol. 58, pp. 203-14.

Singh, K.N. (1975), An approach to the study of the Morphology of the Indian Village, in R.L. Singh, K.N. Singh & Rana P.B. Singh (eds.), *Reading in Rural Settlement Geography*, NGSI Publication, Varanasi, pp. 425-36.

Singh, K.N. (1975), Social Patterns and Space Articulation in the Indian Village, *National Geographical Journal of India (NGJI)*, Vol. 21 (3 & 4), pp. 220-230.

Singh, K.N. & D.N. Singh (eds.) (1985), *Rural Development in India*, NGSI Publication Varanasi.

17

URBAN SETTLEMENTS

Urban settlements are basically the result of non-agricultural activities. But it has not been an easy task to define an urban centre. According to the definition of the Census of India, a human settlement is defined as 'urban' if it has (a) a minimum population of 5,000 persons; (b) 75 per cent of the male working population engaged in non-agricultural activities; (c) a density of at least 1,000 persons per km^2, and (d) distinct urban characteristics (e.g. industrial areas, areas of tourist importance etc.). A settlement which has a municipality, corporation, cantonment or notified areas is also classified as a town. Anomalies arise concerning the population size of villages with population between 5,000 to 10,000 and above, and towns with population between 5,000 to 20,000 in the whole settlement system. These are the rurban settlements. In this group perhaps the functional structure is more important than the population size. It can be found that some of the accessible large villages (5,000 to 10,000 persons) are more urbanised than some of the equally accessible small towns.

Urban Population

In India the urban population has steadily increased from 10 per cent in 1901 to 23.76 in 1981, partly owing to migration. The annual growth rate of urban population during 1901-81 was 4.6 per cent. In absolute numbers, urban population increased by 84 million. The early years of this period was characterised by slow growth or decline in the proportion of urban to total population which was partly due to natural disasters and partly due to the very slow pace of industrial and economic development. The cause of massive growth in the decade 1941-51 was the heavy inflow of refugees to the urban centres after the partition of the country in 1947. The abnormal inflow of refugees accounted for 6.2 per cent of the urban growth, thus yielding a rate of nearly 35 per cent increase in urban population. The 1951-61 decade was marked by rapid strides in industrialisation and it was generally expected that urbanisation too should be rapid during this decade, but the 1961 data do not give any indication of accelerating urbanisation. It may be due to the relative slow pace of economic development in India. The evidence seems to indicate that rate of development was less in comparison to most of other countries roughly at similar stages. The root cause of high growth rate in 1961-71 and 1971-81 due to the inability of the agricultural sector to absorb the rapidly increasing rural population. Due to inadequate job opportunities in the countryside people migrate to cities and towns in search of employment. Since the industrialisation has largely taken place in and around cities and towns, the rural population shifted to cities. A total of 15.7 million persons have migrated from rural to urban areas during the decade 1971-81.

Growth and Size of Urban Centres

In the past eight decades the urban population has increased six folds while the number of towns of various sizes putting altogether could not even double during this period. In 1901 there are only 1,815 towns in the country but by 1951 this figure shot up to 2,818, while in 1961 due to changed definition of the urban areas the number came down to 2,311 (Table 17.1). After this change the same definition is continuing. The Census of India, 1981 has recorded the total number of towns and urban agglomerations as 3,301 (excluding Assam).

The degree of imbalance in terms of high population concentration in cities and multiplication of cities, and the nearstatic proportions in class II and III towns accompanied by a decline in small towns are important factor in Indian urbanisation. The large cities today were not large to begin with but have steadily grown over years. There were only 25 class I cities (1,00,000 population and above) with 22.02 per cent of the total urban population in 1901. In 1981, there are 216 class I towns sharing 60.37 per cent of the total urban population (Map 17.1). As against this the class II and class III towns have maintained more or less same proportion of population in 1901 (11.40% and 16.95%) and in 1981 (11.65% and 14.35%). In case of small towns (less than 20,000 persons, the proportion of population has declined continuously and significantly (Table 17.1).

TABLE 17.1: Number of towns and percentage of urban population, 1901-81

Year	No. of Towns	Class I		Class II		Class III		Class IV		Class V		Class VI	
		No. of Towns	% of Urban Population	No. of Towns	% of Urban Population	No. of Towns	% of Urban Population	No. of Towns	% of Urban Population	No. of Towns	% of Urban Population	No. of Towns	% of Urban Population
1901	1,815	25	26.32	43	11.22	131	15.68	391	20.70	744	20.02	479	6.06
1911	1,801	24	27.68	40	11.40	136	16.35	364	19.53	713	19.29	522	6.75
1921	1,927	30	24.93	45	10.29	146	15.88	370	18.11	740	18.65	592	7.14
1931	2,050	36	31.43	56	11.54	184	16.77	436	17.89	800	11.04	531	5.33
1941	2,226	50	38.37	75	11.46	242	16.21	501	15.72	922	14.97	430	3.27
1951	2,818	77	44.71	92	10.02	327	15.60	611	13.60	1130	12.94	581	3.13
1961	2,311	104	51.54	129	11.15	438	16.84	723	12.75	715	6.85	202	0.87
1971	2,519	150	57.33	173	10.83	561	15.96	830	10.90	638	4.50	167	0.48
1981	3,301	218	60.46	270	11.54	743	14.30	1,059	9.52	758	3.64	253	0.54

Notes:
1. An urban agglomeration has been classified according to the total population.
2. Above data exclude Assam.

Source: A Handbook of Population Statistics, 1988.

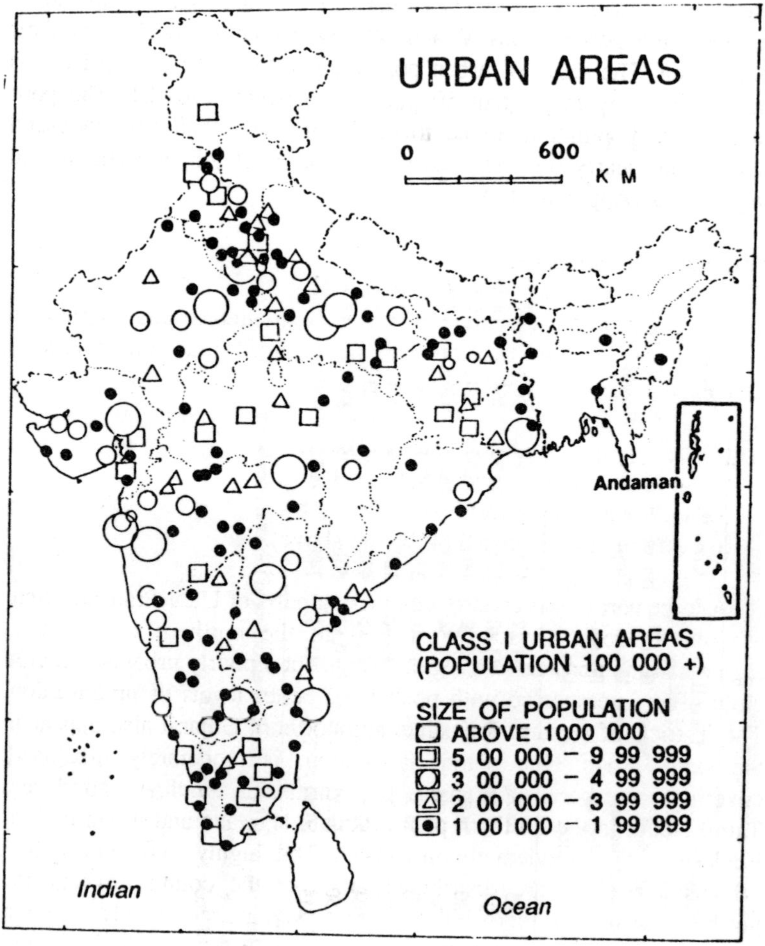

MAP 17.1: Urban Areas

The larger towns have generally grown at a faster rate than the smaller towns. The highest increase of 1,593.98 per cent in urban population is recorded by class I cities for the period 1901-81, while the class II towns recorded 531.21 per cent increase in their population. In the same period class V and VI towns marked decline in their population. The share of urban population residing in class VI towns has declined by more than 50 per cent between 1901-81. The poor increase of population in the towns having less than fifty thousand population clearly indicates the concentration of job opportunities at large towns (Appendix XI).

Patterns of Urbanisation

On the basis of the percentage of urban population to total population in India, the various districts can be grouped into four levels of urbanisation:

(a) poorly urbanised (less than 15.0%)
(b) moderately urbanised (15.0 - 29.9%)
(c) highly urbanised (30.0 - 44.9%)
(d) very highly urbanised (45% and above).

A large part of the country comprising parts of Uttar Pradesh, Bihar Madhya Pradesh, Orissa, West Bengal, the northern mountainous region and the northeastern region brings out a poorly urbanised region quite often interspersed with pockets of better levels of urbanisation. The districts of western Rajasthan and northern Gujarat also appear in the same group. The peninsular region, is moderately urbanised, covering some parts of Maharashtra, Karnataka, Andhra Pradesh and Tamil Nadu. In the northern plain, Punjab, Haryana and western Uttar Pradesh is also moderately urbanised. The highly and very highly urbanised pockets are distributed all over the country except the northeastern and northern hilly and mountainous regions. The growth of large cities due to industrial development are responsible for highly and very highly urbanised pockets. Out of the total districts of various states, 84.62 per cent in Orissa, 83.87 per cent in Bihar, 78.57 per cent in Jammu and Kashmir, 75.00 per cent in Himachal Pradesh, 66.67 per cent in Tripura, 66.66 per cent in Manipur, 57.14 per cent in Nagaland, 56.25 per cent in West Bengal, 55.35 per cent in Uttar Pradesh and 53.33 per cent in Madhya Pradesh are poorly urbanised districts. The percentages of moderately urbanised districts in the states of Haryana (75.00), Karnataka (73.68), Punjab (66.67) and Maharashtra (57.69) are notable. The states of Gujarat (26.32), Punjab (25.00) and Sikkim (25.00) have comparatively large pockets of highly urbanised districts. The very highly urbanised districts of the country are found in Tamil Nadu, Madhya Pradesh, Jammu and Kashmir, Uttar Pradesh, West Bengal, Maharashtra, Bihar, Gujarat and Karnataka. The location of metropolises and large cities have obviously played an important role in transforming the concerned districts as very highly urbanised. The districts of Hyderabad, Greater Bombay, Madras and Calcutta are completely urban, whereas the districts of the Dangs in Gujarat, Lahul and Spiti and Kinnaur in Himachal Pradesh and Waynad in Kerala and Phek in Nagaland are not at all urbanised because of their inaccessible locations and unhospitable terrain.

Million Cities

With the increase in population in bigger cities the number of class I cities has also rose significantly along with million plus cities. In 1901, Calcutta was the only million plus city in the country. In 1911 Bombay joined Calcutta. By 1951, Delhi, Madras, Hyderabad had a million residents each. In the next thirty years, Ahmedabad (1961), Bangalore (1961), Kanpur (1971), Poona (1971), Nagpur, Lucknow and Jaipur (1981) joined the million group. These twelve million plus cities now account for more than a quarter of the nation's urban population. In 1901, they shared 18.6 per cent of the total urban population. However, the million cities recorded more than eightfold increase in absolute population during the period 1901-81. The partition of the country has lead to the mushroom growth of population in the metropolitan cities of the country. After the Independence the decade 1961-71 and 1971-81 have been marked with reasonably high growth rates due to overall development in economy initiated in Five Year Plans. The concentration of population in nine out of twelve million cities may be due to their status as capital city of the state.

Inter-regional Variations

Sharp contrasts in natural and economic landscape lead to sharp inter-regional differences in urban patterns. It must be mentioned here that variations in the industrial resources base and accessibility to transport and power lead to sharper regional urban variations.

The degree of imbalance that is created in the urban and regional system by a metropolis of first order becomes evident if the dominance of the metropolis in its region or state is examined. The trends in metropolitan Bombay for example confirm to core-periphery dichotomy. If the twin city project take full shape, the Bombay metropolis should have 12 million population by 1991. Strictly speaking the Bombay metropolis should be treated as a part of the emerging megalopolis with Bombay and Ahmedabad as the primary node and Pune, Surat and Vadodara as the secondary nodes. The level of economic concentration in Greater Bombay becomes evidence from the following facts.

The metropolitan regions of Bombay and Pune together captured 97 per cent of the state's net increase in factory workers during past 20 years. More than 68 per cent industrial units concentrated in the Bombay city area, comprises key industries like electronic, engineering and chemicals. While stagnating and even declining cotton textiles and agro-processing industries are located in the low production peripheral regions. When there is some resource potential as in the forest and mineral rich areas (Nagpur and Chandrapur) and fishery areas (Konkan), their level of utilisation is low. The rural landscape of the peripheral areas as in the upper valleys of the Tapti, Godavari and Krishna, is characterised by distinct zonation between valley plains with concentration of agriculture and the adjoining uplands and inaccessible watersheds with grasslands, shepherds, nomads and tribals. The limited natural resource endowment and intro-regional inaccessibility and dominance of Bombay all have truncated the urban system.

There are also sharp functional and spatial gaps in urban and market infrastructure in urban industrial pattern associated with industrial development mainly rooted in the regions agricultural and industrial resources and well developed power and transport network. The urban industrial pattern is well represented by Coimbatore district of Tamil Nadu. The major thrust for urbanisation in this region came from industrialisation rooted in the region's resource base. This is the zone of maximum concentration of industrial and market centres (Table 17.2). The population size class of urban centres varies from small towns to class I cities. The region's urban industrial cities function as a part of the country's industrial and market network and contribute their share in the development of national economic space.

In contrast at the intra-regional level, the disparity between the axial belt and the northern (irrigated agriculture and southern regions dry farming and livestock farming) is sharp. This is attributed to the absence of regional planning.

At the micro-regional scale the Ganga plains, the agricultural heartland of India with its high rural population densities and well developed transport and power network is highly urbanised and metropolitanised. Unlike the agriculturally poor inter-metropolitan troughs between Delhi and Bombay and Madras, the inter metropolitan through between Delhi and Calcutta is mainly agriculturally rich and is also a city-rich region. In contrast to the Gànga plains, the core areas of the Mahanadi, Godavari, Krishna and Cauvery deltas have a different urban pattern and are less urbanised mainly because of regional inaccessibility to transportation lines. Punjab is an example of agriculturally highly productive and prosperous state which is characterised by small and medium sized cities and towns.

TABLE 17.2: Growth of population in selected urban and industrial centres

Urban centres	State	Population 1981	Growth rate	
			1961-71	1971-81
Ranchi	Bihar	500.593	82.21	95.89
Durg-Bhilai	Madhya Pradesh	490.158	83.99	99.96
Aurangabad	Maharashtra	316.244	69.14	91.37
Ghaziabad	Uttar Pradesh	291.955	81.29	128.63
Bokaro Steel City	Bihar	261.240	–	143.79
Bhubaneswar	Orissa	219.419	176.07	108.00
Bhatinda	Punjab	127.450	25.00	95.12
Raniganj	West Bengal	119.322	71.74	156.40
Ondal	West Bengal	108.647	51.96	246.62

Source: Census Mapping Survey, 1984, p. 94.

The least urbanised areas are the inaccessible and hill forest areas but even at low urban level, there are intra-regional variations. The sharp intra-regional variations are related to the density of forest cover, accessibility to transport and type of regional economy and state's specific region-oriented development policies.

Like most of the developing countries in India, the towns and cities are growing at a fast pace. Unfortunately, it could not have generated the socio-economic output at desired level. The rural push on account of subsistence agriculture and lack in freedom of choice of economic activities prevailing in the countryside are more responsible for rapid urbanisation.

Morphology of Indian Cities

Unlike other British Colonies particularly in Africa, towns and cities did exist in India before the advent of Europeans. However, European also developed new settlements in the Indian soil. They were the hill stations, administrative posts, railway towns, and garrison towns. These settlements in due course became integrated with the existing settlement fabric and became sizeable as well. But, in any case, these settlements were urban in character. They had most of the urban facilities and connected with road, railways and sometime also with the airways.

The modal of the morphology of the Indian towns and cities is different from the above. The original nuclei of the urban settlements were temple, fort, palace, permanent source of water or a place of some importance. *Chowk*, which does not exactly fit in the western concept of Central Business District (CBD) or down town, has been an essential part of such centres. Business functions are dealt in this area. It is also used for residential purposes. In the wall towns, this zone generally within the surround wall of the old city. The examples are Aligarh, Ahmedabad, Delhi and Mathura. Now these walls are pulled down and circular roads have been constructed on the same alignment. However, in most of the cities, the old walled city gates do exist till now (Fig. 17.2).

With the growth of population, the demand for additional space for accommodation became necessary. Hence, new residential colonies grew up around the old city. These colonies were better planned and spacious. The colonial administration, made certain changes and additions to the earlier urban centres. The introduction of railways from the mid of the last century was a common feature. The railway colony, railway club and institute developed around the railway station. Further; the railway line became a dividing line between the old city and the newly developed area by the colonial administration. Civil lines, courts, jail, cantonment and institutes were on the other side of the railway line. With the passage of time both the settlements became integrated and formed one urban complex.

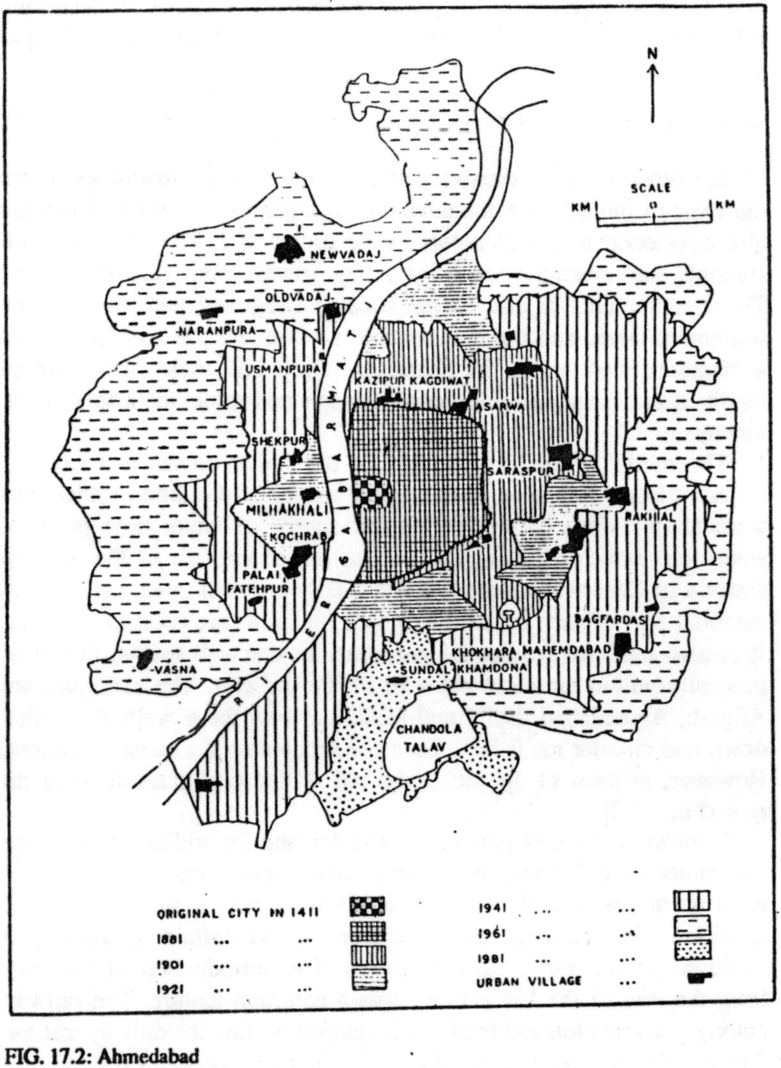

FIG. 17.2: Ahmedabad

The Independence brought certain changes as well. The urban settlements became sizeable due to natural growth and migration. The partition of the country in 1947, rural unemployment and social conflicts, industrialisation, natural and political calamities and the like contributed to migration to the urban areas. This influx required more space for residence and for social facilities. Further, in past 25 years, several campuses, industrial estates, research institutions, universities etc. have come up around the main cities. This campuses are being identified as census towns or outgrowths. They have become large and the areas between these campuses are also built up by residential colonies. The examples of such cities are Bangalore, Kanpur and Ahmedabad.

The merging of census towns around the main city is not only an indication of expanding cities. The cities are themselves merging and forming urban corridors, such as Lucknow-Kanpur, Ahmedabad-Vadodara, around Delhi, Calcutta-Burdwan and the like. Fortunately, the economy is supporting such a development in a predominantly rural country, probably at the expense of small towns.

REFERENCES

Alam, S. Manzoor and V.V. Pokshishevsky (eds.) (1973), *Urbanisation in Developing Countries*, Osmania University, Hyderabad.

Bhattacharya, B. (1985), Urban design and Indian experience in N. Prasad, S. Banerjee and G.K. Dutt (eds.), *Modern Geographical Concept*, Prof. M.R. Chaudhuri Felicitation Volume, Burdwan, pp. 33-22.

Bose, A. (1980), *India's Urbanisation, 1901-2000*, second edition, Tata, McGraw-Hill, New Delhi.

Dube, K.K. and A.K. Singh (ed.), (1988), *Urban Environment in India*, Inter-India Publications, New Delhi.

Manzoor Alam, S. and Fatima Ali Khan (eds.) (1987), *Poverty in Metropolitan Cities*, Concept Publishing Company, New Delhi.

Mitra, Asok *et al.* (1980), *Population and Area of Cities, Towns and Urban Agglomeration*, 1872-1971, Allied Publishers, Bombay.

Munshi, S.K. (1975), The nature of India urbanisation : A review, *Geographical Review of India*, Vol. XXXVII, pp. 287-99.

Nag, P. (1984), India, in P. Nag (ed), *Census Mapping Survey*, IGU commission on Population Geography, Concept Publishing Company, New Delhi.

Prakasa Rao, V.L.S. (1983), *Urbaniszation in India Spatial Dimensions*, Concept Publishing Company, New Delhi.

Rana, P. and G. Krishan (1981), Growth of medium-sized towns of India,, *Geo Journal*, Vol. 5, pp. 33-4.

Ramachandran, R. (1989), *Urbanisation and urban System in India*, Oxford University Press, New Delhi.

Roy, B.K. (1986), Recent urbanisation in India: A scenario, NGJI, Vol. 32(1), pp. 16-29.
Singh, R.L. and Rana P.B. Singh (eds.) (1979), *Place of small Towns in India*, NGSI Publication, Varanasi.
Tanija, K.L. (1971), *Morphology of Indian Cities*, NGSI Publication, Varanasi.
Tewari, V.K., Jay A. Weinstein and V.L.S. Prakasa Rao (eds.), *Indian Cities: Ecological perspectives*, Concept Publishing Company, New Delhi.
Turner, R. (ed.) (1962), *India's Urban Future*, University of California, Berkeley.
Sengupta, S. (1986), Demarcation of residential suburbs of Ahmedabad City, *NGJI*, Vol. 32(1), pp. 49-60.

18

EDUCATION

India was an important educational and cultural centre in the past. India had, Taxila and Nalanda, two very renowned universities in the ancient times. But today our educational system is less of a match in comparison to other developed countries. The literacy rate is low and there is no proper planning for the systematic education all over the country (Table 18.1). The educational system is broadly divided into three sections, i.e. primary, secondary and college, university education.

TABLE 18.1: Per cent distribution of literates by sex and highest educational levels and above, 1971 and 1981, India

Highest Educational level completed	Persons		Males		Female	
	1971	1981	1971	1981	1971	1981
1	2	3	4	5	6	7
Total literates	100.0	100.0	100.0	100.0	100.0	100.0
Less than primary	32.7	30.7	31.2	29.4	36.6	33.4
Primary	35.8	31.5	34.6	30.3	38.2	34.0
Middle	18.0	17.6	18.7	18.0	15.6	16.8
Matriculation or higher secondary	15.5	16.3	13.1	17.9	8.2	12.8
Graduate and above	2.0	3.9	2.4	4.4	1.4	3.0

Primary Education

After Independence, attention was paid to education as a part of national reconstruction initiatives. But hitherto primary education in this country has been the most neglected part of the educational system. There are about 550 thousand primary schools in India. Though government has initiated a scheme of having primary schools more of less covering every village (557 thousand) of the country, but the conditions of those schools are pathetic. The administration of primary education is at present in the hands of local authorities. They maintain their own schools, appoint their teachers, prescribe the courses and textbooks. They also give grants to privately managed schools over which they have little control. There are also privately managed schools which receive no aid from any quarter, their only source of income being the tuition fee paid by their pupils.

Most of the primary schools in the country, mainly in rural areas, are having single teacher only and in few cases there are no teachers. In general, the medium of instruction in primary schools in rural India is the local language. Another important aspect is that only 28.3 per cent of the population in the age group of 5-9 attend school while 69.7 percent is still illiterate and 0.7 per cent is literate but do not attend any school. The attendance in primary school varies between various states and also between rural (0.7%) and urban (46.6%) areas. One peculiar characteristic of primary curriculum is that at present it is more or less a miniature replica of the secondary courses, narrowly theoretical and bookish as the latter. One of the main causes of the failure of primary education system in the country is poverty. In the present economic conditions of the people whatever little money that a child earns by doing some job is a welcome addition to the family income. So most of them have no desire for education nor do they feel any inclination for sending their children to school unless they are compelled to do so.

Secondary Education

The origin of secondary education in India can be traced to the efforts of Christian missionaries and certain nationalists during the late eighteenth and early nineteenth century. After the attainment of Independence, the entire system of education was thoroughly reviewed. A new organisational pattern of secondary education was formulated (1952) as follows:

(a) A middle of junior secondary stage over a period of three years. This stage combined with the earlier primary stage should provide eight years of integrated elementary (basic) education up to age of 14 — the age limit fixed for universal education by the constitution.

(b) A higher secondary stage of three years. This stage should offer a diversity of educational programme to suit the varying aptitudes interests and abilities of children.

(c) Beyond the higher secondary school a three-year degree course or various vocational and technical education was proposed. Hence this proposal had abolished 10 plus 2 plus 3 educational system which was present during British period and started 8 plus 3 plus 3 system. But it is interesting to note that during 1980s educational system in the country has changed again. In the new system, secondary education has been extended for 10 years and higher secondary for 2 years and three year degree course or 4/5 years technical, medical or other specialised courses. For the diversified courses in the higher secondary stage there are seven broad groups or streams : (i) humanities, (ii) sciences, (iii) technical subjects, (iv) commercial subjects, (v) agricultural subjects, (vi) fine arts, and (vii) home sciences.

The idea of universal education up to age of 14 years has been accepted as a goal, but making it a reality is far from being accomplished (Table 18.2). Map 18.1 indicates wide disparities among different regions in attaining literacy levels. The percentage of literates vary from below 30 in Uttar Pradesh or Bihar to above 70 in Kerala. In India there is a problem of equalisation of educational opportunities. Wide differences in the home environment, parents educational and economic status and very uneven standards of educational institutions have combined to produce gross inequalities in educational opportunities. The disparities between the education of boys and girls at all levels, and the level of educational development of advanced and backward classes are more apparent.

TABLE 18.2: Total population literates and illiterates, 1901-81, India excludes Assam

(in million)

Year	Population	Literates	Illiterates	Literacy rate
1	2	3	4	5
1901	235.1	12.6	222.5	5.3
1911	248.2	14.7	233.5	5.9
1921	246.7	17.7	229.0	7.2
1931	273.4	26.0	247.4	9.5
1941	312.0	50.2	261.8	16.1
1951	353.1	58.9	294.2	16.7
1961	428.1	102.6	325.5	24.0
1971	533.5	157.3	376.2	29.5
1981	665.3	241.0	424.3	36.2

Note: Figures up to 1941 are for undivided India.

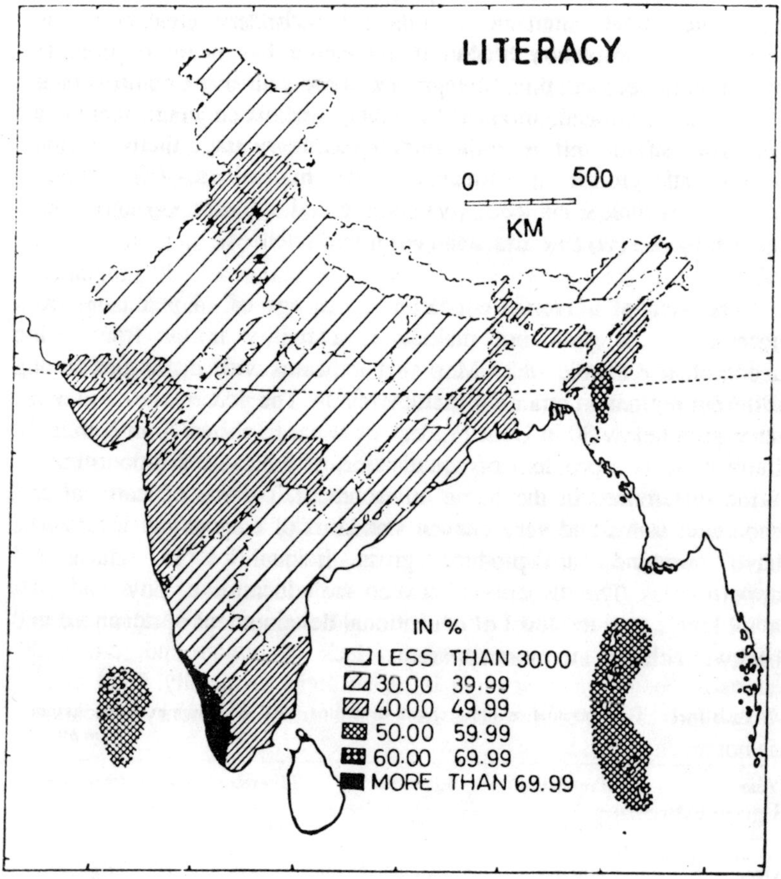

MAP 18.1: Literacy

Another problem which concerns the secondary and higher secondary stages is the medium of instruction. Except in the private schools in the urban areas, secondary education is now given through the medium of the regional language. By and large, it is accepted that English should be the second language at the secondary stage. In the later stage, three language formula for secondary education was proposed. The intention behind this formula has been to promote national unity and emotional integration. This is always a controversial subject. After the declaration of Hindi as an official language, there has been state of turmoil in non-Hindi speaking areas, mainly in the southern states. But, if Hindi is to become an effective official language, then it must be understood by the every person in the country. It is comparatively unimportant whether one state decides to make English compulsory and another gets away with it altogether. Nevertheless, every student in India must learn an Indian language which will be the language of the whole country.

There are various other drawbacks in the secondary education in India. The existing curriculum in schools are overloaded, too bookish and unrealistic. Most of the textbooks are shoddy, badly printed and often full of inaccuracies. The system of examination is unsatisfactory. External examinations at the end of the secondary school stage differ from state to state. In many schools the teachers are often underpaid and overworked and their calibre is mediocre. A large number of teachers take up private tuitions to support their families. This leads to undesirable practices and lack of initiative and interest in teaching in schools. Other than these problems, lack of playground, crowded classes are common in schools in urban areas. Ironically, rural areas where there is so much space, the expansion of secondary education has not reached yet.

Higher Education

After the Independence and the establishment of Planning Commission, planned industrial and agricultural development had become an integral part of the national policy. Hence the demand for trained and technical personnel had increased suddenly. A large number of vocational training and polytechnic schools and colleges were established along with the engineering, medical and agricultural universities. There are 120 universities, 23 deemed universities and 6,040 colleges, 10 institutions of national importance and 26 agricultural universities in India in 1988. These institutions enrol 3.58 million students for graduate, post-graduate and research degrees and diplomas and employ 234 thousand teachers for teaching and research.

The universities of Calcutta, Bombay and Madras were established in 1857. They were modelled on the University of London, that would affiliate colleges, examine students, set standard for courses, prescribe texts and confer degrees. In 1881, the University of Punjab (now in Pakistan) was set up, followed by the University of Allahabad in 1887 to look after the needs of north and central India. In 1915, the Banaras Hindu University was established as a teaching and residential university without the power of affiliation.

At present, there are 182 university level institutions. Of these 131 are universities and deemed universities. There are 48 professional universities as well. Of these, 26 provide education in agriculture and veterinary sciences, 14 in engineering and technology, 7 in medicines and 1 in law. Others provide education in all discipline including some professional disciplines. Out of 6,040 colleges, 1,272 provide education in professional discipline like engineering agriculture, medicine and architecture. Other provide education in arts, physical and social sciences and commerce. Of the 10 institutes of national importance, the five Indian Institute of Technology provide education in technology and science at high level, one provides education in statistics, three in medical sciences and one in Hindi language. (Table 18.3). Besides, there are four Institutes of Management at Ahmedabad, Bangalore, Calcutta and Lucknow which provide advanced management education. The largest number of universities in India belong to the affiliating and teaching type, in which university departments impart instructions at the post-graduate level and undertake research. These universities have a large number of degree colleges affiliated to them providing undergraduate education. Some colleges also have post-graduate teaching and research facilities. A second type comprises the unitary universities like Aligarh, Banaras, Lucknow, Annamalai and Jadavpur. These universities do not have major responsibilities in respect of affiliated colleges. In most of them there is undergraduate and post-graduate teaching as well as research. There are some universities which are in some sense a mixture of the two types. Their territorial jurisdiction is usually confined to the city in which the university is located. Delhi University is a typical example of this type.

TABLE 18.3: Distribution of universities

States/Union Territories	Universities	Management	Technical	Medical	Agriculture Universities	Women	Other	Total
Andhra Pradesh	8		1		1	1	5	16
Arunachal Pradesh	1		—		—		—	1
Assam	2		—		1		—	3
Bihar	6		1		2		2	11
Goa	1		—		—		—	1
Gujarat	7	1	1		1		4	13
Haryana	2		—		2		—	4
Himachal Pradesh	1		—		2		—	3
Jammu & Kashmir	2		—		4		—	3
Karnataka	6	1	—		2		3	12
Kerala	4		—		1		—1	6
Madhya Pradesh	9		—		1		—	10
Maharashtra	6		1		4	1	—	12
Manipur	1		—		—		—	1
Meghalaya	1		—		—		—	1
Orissa	3		—		1		1	5
Pondicherry	1		—		—		—	1
Punjab	2		—		2		—	4
Rajasthan	5		—		1	1	1	9
Tamil Nadu	8		2	1	1	3	3	16
Tripura	1		—	—	—	—	—	1
Uttar Pradesh	13	1	1	1	3	1	6	25
West Bengal	7	1	2		1	—	5	16
Chandigarh	1		0	1	—		—	2
Delhi	2		1	1	1		5	10
Total	100	4	10	4	28	5	36	182

Source: University Handbook, 1988-89

Other Institutions

A new type of institution in higher education is provided by the agricultural universities, established on the American pattern of Land Grant Universities with stress on research and extension work. The first of this kind is Govind Ballabh Pant University of Agriculture and Technology in Pantnagar in Uttar Pradesh in 1961. A substantial part of the funding of the agricultural universities come from Indian Council for Agricultural Research. Some institutions of higher education have been recognised by the Central Government as deemed universities. These institutes generally aim to strengthen their activities in the field of specialisation rather than to make effort towards growing into multi-faculty university of general types. There are various important bodies like University Grant Commission, Council for Scientific and Industrial Research, India Council for Agricultural Research, Indian Council for Social Science Research, Indian Council for Medical Research and others were established to promote higher education and research in the country.

Recent Developments

In respect of teaching and examinations, some notable experiments have been made in recent years by a few universities. By and large the teaching method continues to be the lecturing methods. In certain institutes new educational technologies are available through satellite, TV and video instructions. A large number of universities in the Hindi-speaking states have switched over to Hindi as the medium of education. This pattern has been followed in one or two other states as well, but many universities continue to have English as the medium of instruction with an option to students for using their own language but in the case of professional courses English continues to be the medium.

An important development is the introduction of non-formal education which concerns with distance learning system as mode of education. There are at present as many as 45 distant education institutions in the country, of these four are open universities. These institutions provides fresh educational opportunities to those who have no privilege of university education. The distant education institutions offer a variety of conventional and non-conventional courses and use multi-media in teaching-learning process.

India has a unique characteristic of having the third largest system of higher education with one of the lowest per student expenditure on higher education in the world. If the number of full time students alone is taken into account, India has perhaps the largest higher education system in the world. In view of the low per capita income of the country, the problem today is how to finance the system adequately and provide higher education to all those who seem to want it. This problem is compounded by rapid expansion of the system, especially by the proliferation of a large number of non-viable institutions of higher education. The issue as to whether the expansion of the system should be checked or the number of aspirants for college education should be contained is as important as the question of providing meaningful education to the students.

The National Educational Policy (1986) has emphasised the availability of highly educated, trained and motivated manpower for dealing with the modernisation and globalisation of the economy. Hence there is a need for setting up and maintaining institutions of excellence at all levels in order to tap the best talent and nurture it with care and competence.

REFERENCES

Aggarwal, J.C. & S.P. Agrawal (1984), *National Policy on Education : Agenda for 2001*, Concept Publishing Company, New Delhi.

Biswas A & S.P. Agrawal (1986), *Development of Education in India*, Concept Publishing Company, New Delhi.

Ministry of Human Resource Development (1986), *National Policy on Education 1986, Programme of Action*, New Delhi.

Mahinder, D.C. Chaudhary & K.R.G. Nair (1981), Education and regional development in India, *Indian Journal of Regional Science*, Vol. XIII (2), pp. 170-180.

Raza, Moonis (ed.) (1986), *Educational Planning: A Long-Term Perspective*, Concept Publishing Company, New Delhi.

Raza, Moonis & Aijazuddin Ahmad (1990), *An Atlas a*, Concept Publishing Company, New Delhi.

Reddy, Udaya Bhaskara (1985), Regional disparities in educational development in India : An inter-state Analysis, *Geographical Review of India*, Vol. XLVII (2) pp. 22-27.

Tiwari, A.K. (1984-85), Geography in India Universities : Past, present and future — A review, *The Indian Journal of Geography*, Vol. XIV, pp. 85-94.

19

INDUSTRIAL DEVELOPMENT

After Independence, the Indian Government was at a state where the infrastructure was grossly inadequate. In order to circumvent the constraint in the economy at a national scale covering all aspects of economic and social life, and to initiate simultaneously a process of all round balanced development, it was intended that the traditional society should gradually be transformed into a modern society with the help of a proper planning. Government of India decided to follow neither a full free capitalist path nor a fully free socialist path. It opted for developing a mixed economy in India. Both private and public sectors are being developed side by side.

Industrial Development Through National Plans

Government of India had taken various measures in different plan periods for higher productivity, higher employment generation, correction of regional imbalance and the like. All these broad socio-economic objectives were accepted throughout the three decades of planning in India. Still, slow growth rate, high unemployment, rural-urban disparities and concentration of assets in the hands of big business houses are common.

In the First Five Year Plan industry got less importance than agriculture, irrigation and power. Main emphasis was laid on fuller utilisation of existing capacity. The cotton textiles, sugar, vegetable oils, soda ash, rayon, bicycles etc. had more or less reached the target of production. It is interesting to note, that during this period newsprint, calcium carbide, penicillin, DDT, automatic looms, steel wire, rope, jute spinning frames, deep well turbine pumps and motors and transformers of higher ratings were first manufactured in India.

In the Second plan period, main emphasis was to lay firm foundation of industrial progress. So, the main priorities were the increased production of iron and steel, heavy chemicals, heavy engineering and machine building industries, expansion of aluminium, cement, chemical dyestuff, phosphate fertiliser etc. Industrial progress was most remarkable during this plan period. New steel plants were set up in public sectors. In chemical industries the advance was on wide front. The index of industrial production rose from 100 in 1950-51 to 194 in 1960-61. Many industries were developed for the first time during this period.

In the Third plan period main emphasis was given on machine building programmes and also on the development of the related skills, technical know-how and designing capacity. The Third Plan period was followed by annual plans for three years, and during this period the industrial development has been uneven. During the second half of this period, the economy was not very stable, hence the industrial production has also declined.

In the Fourth plan period economy of the country was recovering, so main thrust in this period was on accelerating industrial growth by better utilisation of existing capacity. The Planning Commission has given emphasis in the industrial development in private sector. However, the important industries like steel and fertilisers production have fallen short of target due to operational problems.

In the Fifth Plan period production of export-oriented goods and articles of mass consumption got emphasis. The sharp increase in the prices of foodgrains, fertilisers and oils has upset the planning of this period. This plan was terminated one year earlier. During this period the volume of production has expanded both in size and spectrum. Due to the growth of indigenous skill the foreign assistance was less in this period.

The industrial policy during the Sixth plan aimed at optimum utilisation of existing capacity. Production target were achieved in industries like petroleum, machine tools, passenger cars, motor cycle and scooters, TV receivers etc. During this period 500 MW power generation units were commissioned, manufacture of 500 MW turbo generator and boiler had started and, fuel efficient 800 cc passenger cars came into market. In the Seventh Plan period 6.3 per cent growth rate achieved. The detail results and achievement of this plan are awaited.

During the different five year plan periods the expansion and diversification of India's industrial structure helped to transform the economic scenario in the country. The country has attained self-sufficiency with regard to sugar and cement machinery, power boilers, material handling equipments and a large number of consumer durables. In recent years considerable progress has been recorded by the electronics industry, thermo plastics and petro-chemical industry.

Industrial Situation in India

India has one of the largest industrial bases in the world. It has helped in diversifying the economy and in reducing dependency on agriculture. Industries can be classified into several groups, such textiles, metals and metal products, chemicals and chemical products, electricity, food and allied products. They account for 65 per cent of the total gross output originating from the factory sector (Table 19.1)

TABLE 19.1: Production for selected industries, 1987-88

Basic Metals		
1. Steel ingot	18.29	million tonne
2. Pig iron	1.22	million tonne
3. Saleable Steel	15.68	million tonne
4. Alloy & special steel	800	'000 tonnes
5. Aluminum	276.5	'000 tonnes
6. Copper	74.5	'000 tonnes
Non-metallic Mineral		
7. Cement	30.1	million tonne
8. Petroleum products	33.29	million tonne
9. Basic chemicals	1600	'000 tonnes
10. Agricultural Chemicals	5245	'000 tonnes
11. Petro-Chemicals	1510	'000 tonnes
12. Manmade Fibre	173.6	'000 tonnes
13. Cotton Yarn	1392	mi kg
14. Cloth	11950	mi m.
15. Jute	1368	'000 tonnes
16. Drugs	2204	Rs. Crores
17. Paper	1361.20	'000 tonnes
18. Newsprint	197.07	'000 tonnes
19. Railway wagons	16.1	'000 S
20. Commercial vehicles	106.0	'000 S
21. Diesel engines	187	'000 S
22. Fertilisers	7070	'000 tonnes
23. Sugar	8.5	million tonne

Source: Handbook of Industrial Statistics, 1989.

Textile Industries

The textile sectors employ nearly 17 per cent of total workforce. This sector is also leading in account of gross output ex-factory value. The cotton textile is leading in the textile industry in the country. It employs about a million of worker and manufactures about 1.2 tonnes of cotton yarn and 9 thousand million of cotton clothes in the mills, handlooms and powerlooms.

The size of a cotton mill depends primarily on the number of spindles installed. Tamil Nadu appears to have the largest number spindles, some 4 millions, which is corroborated by the fact that this state has the largest number of spinning mills. Maharashtra and Gujarat ranked second and third in the number of spindles installed, though most of cotton mills in these states are of composite type. The cotton textile industry is faced with several major problems, viz. short supply of long-staple cotton in the country, insufficient textile machinery, acute power shortage, high price in the international market, etc.

The cotton mills of India manufacture four varieties of cotton cloths: coarse, medium, fine and super fine. Since the cotton textile industry is partly raw material oriented and partly market oriented, the cotton mills are widespread than any other industrial plant. Bombay, Ahmedabad, Surat, Kanpur, Delhi, Calcutta, Nagpur, Sholapur, Coimbatore and Madurai are the major centres of cotton textile industry.

Jute fibre was known in India in ancient times. In the modern era most of the jute mills of the country numbering a little over 100 are located on the banks of the Hooghly river within easy reach from Calcutta and areas where raw jute used to be assembled from Bangladesh. Three other jute mills are in Assam, Orissa and Tripura. Regarding woolen industry, there is not much scope for development because of the dearth of good quality wool. Most of the larger woolen mills in Punjab and western Uttar Pradesh are facing problems such as high import duty on raw wool and restrictions on capacity expansion of existing woolen mills. Further, natural silk of nearly 3.5 million kg of all varieties are produced in India annually. Karnataka ranks first in the production, followed by West Bengal, Jammu and Kashmir, Tamil Nadu, Assam, Bihar, Madhya Pradesh and Orissa. In addition, India has made a rapid growth in the production of man made fibres and fabrics during last 30 years or so. Rayon is found admirably suited to supplement cotton, to blend with cotton and to be used in pl:

cotton for everyday as well as for ceremonial use.

Food Industry

The food and allied products are the most widely scattered all over the country in 4,000 individual factories. The sugar industry is the second largest in the country. The country manufacture more or less same quantity of sugar as that produced in the U.S.A. Most of the 250 sugar factories of India are located in Uttar Pradesh and Maharashtra. Besides the production of centrifugal sugar, India also produces non-centrifugal sugar, molasses by vacuum pan sugar factories. Like any other industries the sugar industry has its problems which hamper its production. To improve the sugar industry, it is essential to improve direct contact between the factories and the cane grower in order to ensure regular flow of sugar cane to the mills.

Tea is one of the largest foreign exchange earner of India. Assam valley leads in the production of tea. West Bengal ranks second with its prosperous tea gardens in Jalpaiguri (Duars) and in Darjeeling hills. The latter is known for the production of the finest tea so far as flavour is concerned. In south India, tea is produced in Nilgiris, Coimbatore and Madurai districts.

India produces more than one lakh tonne of coffee. Most of the Indian coffee come from Karnataka and rest from Nilgiri hills of Tamil Nadu, and the hilly areas of Kerala. The Coffee Board of India mainly controls the coffee industry in the country.

Vanaspati, a hydrogenated oil was first produced in the country in 1930. Uttar Pradesh is the largest producer of vanaspati, followed by Maharashtra and Delhi. Vegetable oils are the raw materials for the manufacture of vanaspati. The food industry of India also includes edible oil mills, rice and wheat mills dairies, cashewnut processing mills, tobacco factories, fish and fruit canning industry and meat packing industry.

Metal-based Industry

Iron and steel industry rank first among the metal-based industry in India. In 1874, the Bengal Iron and Steel Company started producing pig iron in its blast furnace at Kulti, in West Bengal. Steel ingots were produced in 1907 at Sakchi in Bihar, now named Jamshedpur after the name of Jamshedji Tata. Since then, a number of iron and steel plants have been established in the country. In India the production of pig-iron had increased considerably after Independence. But, India still rank 16th in the world in the production of steel. Further there are at present seven major iron and steel works in India of which four plants have been set up in the public sectors after Independence. All these plants have been set up in the collaboration with the industrialised countries such as U.S.S.R., West Germany and the United Kingdom.

As mentioned earlier, the first modern steel plant, i.e. Tata Iron and Steel Company was set up at Sakchi (Jamshedpur) on the banks of the Subarnarekha river. The location and subsequent growth of this plant is mainly due to the easy availability of rich iron ore, coal, manganese and limestone from the neighbouring areas. The proper management by Tata Group has helped this steel plant to become leading producer of steel in the country. Indian Iron and Steel Company located at Burnpur and Kulti in West Bengal also started producing steel before Independence. This company also has the locational advantage regarding raw materials like TISCO. But due to the poor management the company was almost closed during 1970s. The government has taken over the IISCO and it is functioning as public sector undertaking now. Mysore Iron and Steel Limited, situated at Bhadrawati in Karnataka is the only steel plant located in the southern part of the country and also far from the main coal bearing regions. This plant was also set up prior to the Independence and utilises the huge charcoal reserves of neighbouring regions in absence of metallurgical coal. In recent year, electric power is used as well (Appendix XIII).

After Independence, during First Five Year Plan, the Government of India has decided to established iron and steel plant in public sector. With this objective, Hindustan Steel Ltd. was established. First Iron and Steel plant was established at Rourkela in Orissa. It was followed by at Bhilai in Madhya Pradesh and at Durgapur in West Bengal. All these plants were established during Second Five Year Plan. Bokaro plant in Bihar is comparatively new. All these steel plants are situated in and around mining areas of the country.

Aluminium ranks next to iron and steel in the metallurgical industry in India. The aluminium industry is basically a power intensive one. The importance of aluminium as versatile metal among the group of light metals and hence every attempt is made in the country to develop hydel power and thermal power to feed aluminium manufacturing plant. In India aluminium plant is scattered all over the country. They are in Hirakud (Orissa), Alwaye (Kerala), Belgaum (Karnataka), Renukoot (Uttar Pradesh), Metur (Tamil Nadu), Asansol (West

Bengal), Korba (Madhya Pradesh), and Ratnagiri (Maharashtra).

The production of copper has increased a lot after the Independence. Hindustan Copper Ltd., a Government of India enterprise, is primarily responsible for copper mining, smelting and refining. There are six constituent units, of which three are located in Rajasthan two in Bihar and one in Madhya Pradesh. These units produce copper and alloys like brass in various forms including sheets and strips.

Engineering Industries

Both mechanical and electrical industries are fairly well developed in India. There are more than 1.25 lakh engineering units in India. Some of the major products of the engineering industries are: power generation, transmission and distribution equipments, plant and machinery for steel, chemical and fertiliser plants, electrical and construction machineries, machine tools, railway rolling stock, agricultural equipments etc. Ranchi in Bihar is the main centre for manufacturing heavy mechanical and industrial machinery. Visakhapatnam, in Andhra Pradesh is the principal manufacturing centre of plants for fertiliser and other processing industries and petroleum and other heavy chemical and allied industries. Machinery for heavy electrical industries are principally manufactured in Bhopal, Tiruchchirappally, Hyderabad, Haridwar and Bangalore.

Non-metallic, Chemical and Allied Industries

Due to the rapid industrialisation and urbanisation, the requirement of cement has increased sharply in India. But the production of cement in India is far less than that of most of the developed countries. Since limestone is needed in large quantities, most of the cement mills are located near limestone deposits. The Cement Corporation of India has three large cement factories in Madhya Pradesh, Karnataka and Assam. Besides, there are 55 other cement factories.

There are nearly 150 fertiliser factories in India. The Fertiliser Corporation of India and the National Fertiliser Ltd. were mainly responsible for the production of fertiliser in the country. They have their units at Sindhri (Bihar), Nangal (Punjab), Trombay (Maharashtra), Gorakhpur (Uttar Pradesh), Namrup (Assam), Durgapur (West Bengal) and Alwaye (Kerala). The recently set up Hindustan Fertiliser Corporation has operating units at Barauni, Namrup and

Haldia. In case of insecticides, India mainly produce DDT in its factories at Delhi, Alwaye (Kerala) and Rasayani (Maharashtra).

Drugs and Pharmaceuticals

In order to reduce the import of bulk drugs and pharmaceuticals, several establishments have been set up in the public sector, such as Hindustan Antibiotics Ltd. at Pune for manufacturing penicillin and streptomycin and vitamin C, and Indian Drugs and Pharmaceuticals Ltd. (IDPL) at New Delhi, for manufacturing a variety of antibiotic and drugs. IDPL runs three plants: antibiotic plants at Rishikesh, synthetic drugs plant at Hyderabad, and surgical instrument plant at Madras. Besides, the Hindustan Organic Chemicals Ltd.; Rasayani in Maharashtra manufacture various organic chemical required by drugs, dyestuffs and plastic industries.

Petro-Chemical

Plastics, rayon yarn, and synthetic rubber are some of products of the petrochemical industries. Vadodara (Gujarat) and Bongaigaon (Assam) are the two centres of petro-chemical industry in the public sector (Appendix XIV).

Other Industries

There are more than 120 units which manufacture machine tools. The major public sector understanding known as Hindustan Machine Tools Ltd. has manufacturing units at Bangalore (Karnataka), Pinjore (Haryana), Kalamesery (Kerala), Hyderabad (Andhra Pradesh) and Srinagar (Jammu and Kashmir).

India produce commercial vehicles, including jeeps. The passenger cars, scooters, and motor cycles are manufactured at different parts of the country, viz. Bombay, Bangalore, Calcutta, Delhi, Jamshedpur, Pune, Kanpur, etc. Locomotives were first manufactured at Chittaranjan in West Bengal. There is Diesel Locomotive Works at Varanasi in Uttar Pradesh and Integral Coach Factory at Perambur in Tamil Nadu. There are nearly 100 ship and boat building and repairing shipyards in the country of which Hindustan Shipyard Ltd. takes the foremost place. There are three more ship-building yards at Bombay, Calcutta and Cochin. The Garden Reach workshop at Calcutta also has adequate ship repairing facilities and also produce craft and vessels for the coastal shipping.

There are nearly 120 paper mills in India. The Government of India took initiatives in developing pulp and paper by establishing the Hindustan Paper Corporation Ltd. (Appendix XV).

Industrial Regions

India has five major and fourteen minor industrial regions, according the average daily factory employment (Map 19.1). Though the Hooghly belt was prominent in the country till 1960s but Bombay-Pune and Ahmedabad-Vadodara regions are important now. Among all industrial regions, Chota Nagpur plateau has recorded maximum growth.

Hooghly belt is elongated along the river Hooghly with twin cities Calcutta and Howrah at the centre. Main manufacturing centres are Naihati, Jagatdal, Shamnagar, Tribeni, Belur, Liluah, Andul etc. This region took its origin from the British period. Jute and engineering machineries were some of the industries which thrived here. This region faced great crisis due to partition which has created the shortage of raw jute. Labour unrest, power problem, lack of capital are also responsible for the decline of this region.

Bombay-Pune region flourished as an important cotton textile region in the centres like Bombay, Vile Parle, Thane, Bhandup and Pune. This region also faced a set back after partition for the non availability of raw cotton. But this region is fortunated for having petroleum in and around. Hence, most of the industries have diversified to petro-chemical, synthetic fibre and various other chemical products. Pune has become a major chemical and machinery producing city. This region has had capital availability and less labour trouble. Obviously, diversification was possible.

The Ahmedabad-Vadodara-Surat region is now fast growing industrial region regarding petro-chemical, fertiliser, synthetic fibre, textile and other chemical groups. This region also started cotton textile industry but shifted its focus to chemical due to the availability of natural oil and gas. This region promises a bright future.

The Madurai-Coimbatore-Bangalore region is also famous for cotton textile industry. But, the establishment of an industrial complex at Bangalore like Hindustan Machine Tools, Indian Telephone Industry, Hindustan Aeronautics etc. controlled by the government has further stimulated the growth of industries in this region.

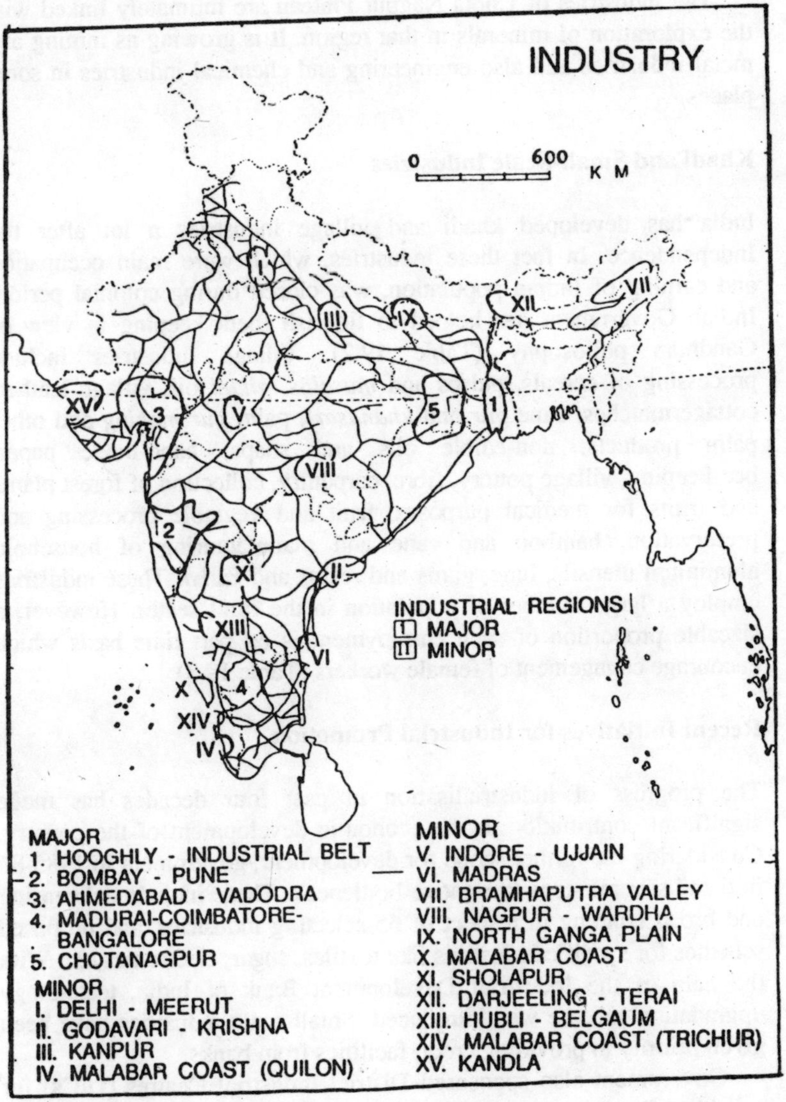

MAP 19.1: Industry

The industries of Chota Nagpur Plateau are intimately linked with the exploration of minerals in that region. It is growing as mining and metal industries and also engineering and chemical industries in some places.

Khadi and Small-Scale Industries

India has developed khadi and village industries a lot after the Independence. In fact these industries, which were main occupation and earning of Indian population, was closed during colonial period. Indian Government has helped to flourish them keeping in view of Gandhian philosophy (Table 19.2). Village industries include processing of cereals, pulses and *masalas, ghani* oil, village leather, cottage matches, cane *gur* and *khandsari,* palm *gur making* and other palm products, non-edible oils and soaps, hand-made paper, bee-keeping, village pottery, fibre, carpentry, collection of forest plants and fruits for medical purposes, fruit and vegtable processing and preservation, bamboo and cane and manufacturing of household aluminium utensils, lime, gums and resins and *katha*. These industries employ a large number of population in the rural sector. However, a sizeable proportion of such employment is on part time basis which encourage engagement of female workers (Table 19.3).

Recent Initiatives for Industrial Promotion

The progress of industrialisation in past four decades has made significant contribution in the economic development of the country. Considering the further scope for development, government in 1985-86 initiated several steps to remove bottleneck. These include delicensing and broad banding in respect of 65 selecting industries, and proposed schemes for specific industries like textiles, sugar, electronics etc. With the help of the Industrial Development Bank of India, technology upgradation scheme was introduced. Small scale industries have been given priority in providing credit facilities from banks.

Government also sponsored District Industrial Centres (DICs) for providing focal point at district level for the promotion of small, tiny village and cottage industries which are widely distributed in rural and semi-urban areas. The aim is to provide essential services and support at pre-investment, investment and post investment levels. It is likely to provide rural employment in industrial sector, thereby reducing pressure on agricultural and checking migration to urban areas. There are 422 such approved DICs in the country (Appendix XII).

Another major aspect which is gaining importance in industrialisation, particularly in the urban areas, is environmental pollution. Several laws have been enacted which insists on having pollution control measures with industries. The location policy has been revised in terms of distance from the urban or populated areas. Environment clearance of projects has become essential. State pollution control boards have been established to monitoring pollution.

TABLE 19.2: Production of khadi and village industries, 1985-86

(in Rs. Lakhs)

State/UT	Khadi	Village industries
1	2	3
Andhra Pradesh	700.22	2679.49
Assam	49.88	399.47
Bihar	873.62	2790.53
Gujarat	841.54	3911.84
Haryana	160.98	1734.79
Himachal Pradesh	67.79	591.64
Jammu and Kashmir	122.83	564.79
Karnataka	618.96	2560.92
Kerala	228.67	1917.51
Madhya Pradesh	168.77	601.12
Maharashtra	247.81	5116.37
Manipur	1.70	296.40
Meghalaya	—	82.26
Nagaland	1.94	22.68
Orissa	39.44	694.06
Punjab	486.07	1715.23
Rajasthan	1138.31	2395.88
Sikkim	1.80	0.65
Tamil Nadu	1078.39	7614.21
Tripura	15.16	268.71
Uttar Pradesh	2565.09	8101.32
West Bengal	638.98	732.37
Chandigarh	0.53	165.32
Dadra & Nagar Haveli	0.76	0.29
Delhi	5.72	.83
Goa, Daman and Diu	—	24.10
Pondicherry	—	17.87

Source: Annual Report, KVIC, 1987.

TABLE 19.3: Production and employment in KVIC, 1987-88

Industries	Production Rs. Crores	Employment (in lakh persons)	
		Full time	Part time
Khadi	106.85	3.95	8.11
Village Industries	27.08	0.21	0.20
P.G. P. Masala			
Ghani Oil	80.20	0.28	0.19
Leather	46.26	0.75	0.24
Cottage match	31.05	0.53	-
Cane *gur* etc.	100.78	1.33	-
Palm *gur* etc.	22.39	-	4.30
N. edible oil etc.	17.25	0.04	1.52
Handmade paper	3.57	0.05	0.01
Bee-keeping	7.59	-	1.71
Village pottery	24.14	0.94	0.60
Fibre	12.07	0.80	0.53
Carpentry etc.	35.88	0.83	0.17
Lime manu	9.74	0.16	0.11
Gobar gas	22.47	0.10	-
Collec. of forest plant	0.99	x	0.15
Manu. gums. & resins	0.09	x	0.02
Manu. of *Katha*	2.64	x	1.72
Fruit & vege. process	1.26	x	0.08
Bamboo & Cane	0.65	x	0.02
Manu. of alu. utensils	4.41	0.26	0.24

x Less than 500

Source: Annual Report KVIC 1988-89.

India's economic policy regarding industry has helped in the industrialisation of the country. At present India is one of the leading industrialised countries among the developing countries. It is very pleasant to note that the Government of India has not neglected the traditional crafts as well. In fact government policy has helped in flourishing all types of industries.

REFERENCES

Chaudhuri, M.R. (1962), *Indian Industries, Development & Location*, Calcutta.

Chaudhuri, M.R. (1978), *The Tea Industry in India*, Calcutta.

Chaudhuri, M.R. (1982), *The Jute Industry in India*, Calcutta.

Chaudhuri, M.R. (1982), *The Public Sector in Indian Industries*, Calcutta.

Chaudhuri, M.R. (1982), *The Public Sector in Indian Industries*, Calcutta.

Devi. D. Radha (1985), Change in the Industrial composition of Indian workers a districtwise analysis, *Demography in India*, Vol. XIII (1 & 2), pp. 141-52.

Kuriyan, G. (1962), Spatial distribution of industry in India, *Indian Geographical Journal*, Vol 6 (1).

Karan, P.P. (1964), Changes in industrial location, *Annals of the Association of American Geographers* Vol. 54, pp. 336-54.

Sengupta, P. (1959), *The Indian Jute Belt*, Orient Longman, Calcutta.

20

TRANSPORT AND COMMUNICATION

India's transport system comprises of roads, railways, inland waterways and airways which are fairly well developed for a third world country. In fact the paved road network of India is estimated to the second largest in the world, next to the U.S.A.; and the Indian railway system is recognised as the largest in Asia. India has about 52,000 km of major rivers and 4,300 km of navigable canals for mechanised craft. India is the second largest shipowning country in Asia and ranks sixteenth in the world.

Roads

Before 1947 in India the road system was developed without reference to the railway systems. In consequence, the majority of the trunk roads ran parallel with the railways while road bridges interfered with water-borne traffic. Indian roads have developed very much after Independence. They are broadly classified into four categories, e.g. national highways (31,769 km), state sector roads (94,680 km), and major district (44,104 km) and other roads (2,79,970 km) depending on the authorities controlling them. The Central Government is directly responsible for the national highways system. Some of these highways like the Grand Trunk Road connecting Calcutta and Peshawar in Pakistan, have existed for a long time. This road is now known as NH2 and serves the needs of the densely populated Ganga plain, over a distance of 1,503 km between Calcutta and Delhi. The longest national highway (NH7) runs from Varanasi in the middle Ganga plain to Kanyakumari at the extreme southern end of the country, a distance of 2,372 km. Delhi is the meeting place of five national highways coming from Calcutta, Bombay, Lucknow, Amritsar and Fazilka, a border town (Map 20.1).

Border roads play an important role in the economic development of the Himalayan border regions and also in the northeastern part of the country. Their construction and maintenance have been entrusted to the Border Roads Organisation. The Manali-Leh Road in the Western Himalaya, passes over a terrain of 4,270 m high for a long distance and crosses four mountain passes of 4,875 – 5,485 m high.

TABLE 20.1: Roads

States/ Union Territories	Total Roads (%)	Highways (%) National	Highways (%) State	Other Roads (%)	Surfaced (%)	Per 100 km^2
Andhra Pradesh	9.43	2.02	4.68	93.3	51.00	42.23
Arunachal Pradesh	0.96	2.79	—	97.21	17.20	14.16
Assam	4.20	4.24	2.95	92.81	11.38	65.96
Bihar	5.24	3.39	6.49	90.12	39.66	37.11
Goa, Daman & Diu	0.62	2.94	3.04	94.02	49.35	199.21
Gujarat	4.05	2.85	18.36	78.79	74.90	25.43
Haryana	1.69	3.13	15.02	81.85	89.98	47.23
Himachal Pradesh	1.48	3.22	17.79	78.99	23.29	32.90
Jammu & Kashmir	0.80	6.00	6.95	87.05	57.57	4.45
Karnataka	7.91	2.03	8.02	89.95	58.14	50.82
Kerala	8.09	0.84	2.08	97.08	20.04	256.23
Madhya Pradesh	5.35	4.18	17.55	78.27	77.30	14.86
Maharashtra	11.65	2.06	13.33	84.61	50.19	46.62
Manipur	0.39	8.93	9.94	81.13	38.61	21.62
Meghalaya	0.40	9.47	—	90.53	54.79	22.04
Mizoram	0.20	9.59	—	90.41	45.24	11.86
Nagaland	0.47	1.81	6.87	91.32	13.66	34.92
Orissa	8.55	1.54	2.69	95.77	12.19	67.71
Punjab	2.65	2.99	5.82	91.19	100.00	64.76
Rajasthan	4.55	4.50	12.98	82.52	62.39	16.36
Sikkim	0.08	6.00	21.00	73.00	100.00	0.79
Tamil Nadu	9.63	1.57	1.53	96.90	79.52	91.20
Tripura	0.55	2.92	2.00	95.08	16.48	64.65
Uttar Pradesh	7.47	2.69	8.71	88.60	59.87	31.08
West Bengal	3.34	3.96	7.65	88.39	42.23	46.37
Andaman & Nicobar Is.	0.05	—	42.0	58.00	94.0	7.82
Chandigarh	0.01	100.00	—	—	100.00	12.28
Dadra & Nagar Haveli	0.02	—	—	100.00	79.67	50.10
Delhi	0.03	27.30	—	72.70	95.40	21.98
Lakshadweep	—	—	—	—	—	—
Pondicherry	0.14	—	1.99	98.01	48.06	367.07

Source: Ministry of Shipping and Transport.

MAP 20.1: National Highways

In general there is the regional or statewise variation in terms length of roads (Table 20.1). The mountainous and hilly regions in the north and east have the least 4–6 km length of surfaced roads for every 100 km². On the other hand, the Indus plain in the north and the Deccan plateau in the south have the greatest length of surfaced roads for every 100 km². These roads are being used by vehicles of different age, size and varieties. They are also used for the transportation of commodities. There were 10 million automobiles on roads. Maharashtra had the highest number of automobiles followed by Uttar Pradesh, Karnataka and Gujarat. In last 15 years (1971-86), the number of vehicles increased over 5 times.

Railways

Railways were introduced in India in 1853 by connecting Bombay with Thane. Within its present territory, it has 61,385 km long routes (Table 20.2) connecting over 7 thousand stations all over the country. Railways serves from coastal region to mountains by running about 11,000 trains everyday. It carried 307 million tonnes of freight and 3,580 million passengers in 1986-87. The earnings from the former was Rs. 4,991 crores and Rs. 1,940 crores from the latter. The neighbouring landlocked countries, Nepal and Bhutan are also dependent on the Indian railways for the transhipment of their goods.

TABLE 20.2: Railways

Zone	Headquarters	Route Distance (km)
Central	Bombay V.T.	6,371
Eastern	Calcutta	4,238
Northern	New Delhi	10,975
North-Eastern	Gorakhpur	5,163
North-East Frontier	Maligaon, Gauhati	3,580
Southern	Madras	6,701
South-Central	Secundrabad	7,023
South-Eastern	Calcutta	7,041
Western	Bombay Churchgate	10,293
	Total	61,385

Note: About 8.5 per cent is electrified.
Source: NATMO Occasional Monograph 2, p. 7.

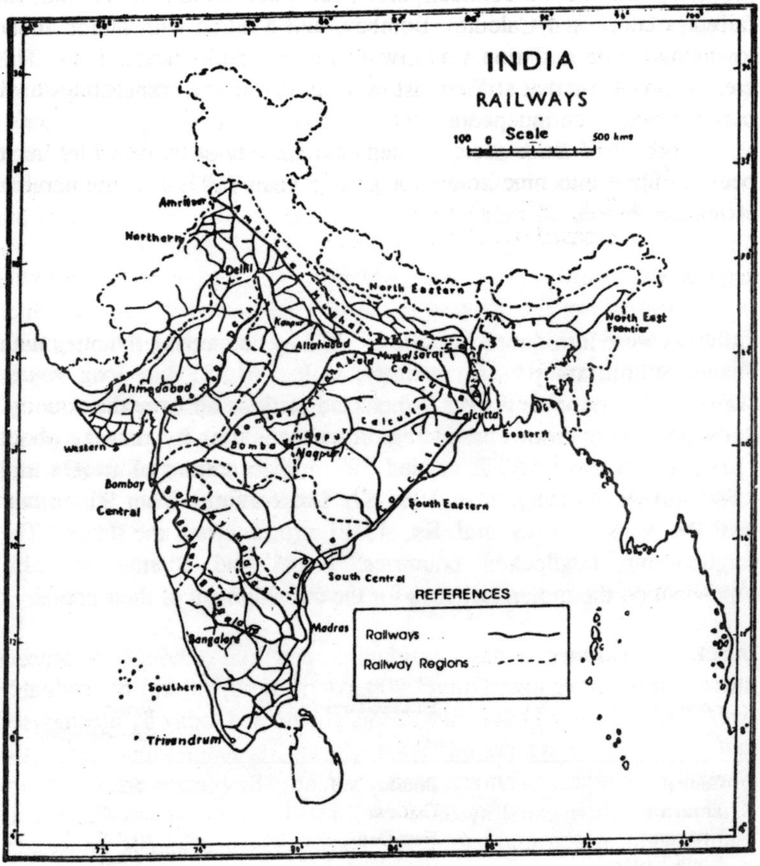

MAP 20.2: Railways

Indian Railways play a crucial role in the building up of infrastructure required for the growing and diversifying the economy of the country. During the British regime railway construction was primarily intended to connect ports with internal markets. At first the railways connected Calcutta, Bombay, and Madras with major inland towns and little attention was given to interconnect feeder lines. This created problems that still persist even though many branch lines have been opened since Independence (Map 20.2).

Before 1947 there were 37 separate railway systems which have been grouped into nine zones for greater economy and administrative efficiency (Table 20.2). The increase in the number of passengers has not been commensurate with the facilities available to the public, and serious overcrowding in the passenger coaches persists. Another problem affecting the railway is the diversity of track gauges, which causes considerable and expensive delays at transhipment points throughout the country.

Mineral fuels especially coal are the principal commodities which constitute more than one-half of the total freight carried by the railways. The second largest item is manufactured goods of which cement, iron and steel, and chemical fertilisers form about 90 per cent of the bulk.

Waterways

Inland water transport has considerable potential although its present utilisation is rather low. Only 1,700 km of the 5,200 km of navigable rivers and 331 km of navigable canals are utilised today by mechanised craft. The waterways comprise Ganga and its main tributaries – the Gandak, Ghaghra, Yamuna and Gomati, Brahmaputra, Godavari, Krishna and their canal systems such as; the Buckingham Canal-1 in Tamil Nadu and Andhra Pradesh; backwaters and canals of Kerala; Tungabhadra canal; and lower Narmada river. Sunderban routes connecting the Ganga delta with Assam (through Bangladesh) also play an important part in the movement of goods.

India has a large merchant ship fleet, perhaps the largest among the developing countries of the world. The country's operative tonnage in 1980 was 5.7 million G.R.T. The share of Indian shipping in the transportation of cargo in India's overseas trade has been continuously increasing since 1951. India has ten major ports : Bombay, Calcutta, Cochin, Kandla, Madras, New Mangalore, Marmagao, Paradip,

Tuticorin and Visakhapatnam. There are another 160 intermediate and minor ports scattered along its 6,100 km long coastline. Each of the ten major ports has its special function. Bombay is the biggest port handling a quarter of the foreign trade of India. Calcutta is the largest terminal port in South Asia. It commands the largest and the richest hinterland. Cochin is one of the finest natural harbours in the east. Kandla has been developed as a major port to relieve the pressure of Bombay and has a free trade zone. Madras is a manmade port. Marmagao in Goa enjoys the second position in total traffic tonnage. Paradip in Orissa is primarily for export of ore of iron. Visakhapatnam on the east coast is the only land-locked and protected port with the largest ship-building yard.

The traffic at minor ports has also increased a lot. Some of the minor ports played a vital role in handling imported foodgrains and releaving pressure on the major ports, besides facilitating internal water transport through coastal shipping. The coastal traffic along the Saurashtra and the Gujarat mainland coast is pretty heavy. It is carried principally through three river ports – Bhavnagar, Bharuch and Surat. Coastal streamers run regularly from Bombay port to Mangalore, Cochin and Tuticorin in the south and Surat, Porbandar and Kandla in the north. The cargo carried by coastal shipping consists entirely of bulk commodities such as coal, salt, cement and general cargo comprising foodgrains, fertilisers, jute, tea, iron and steel products.

Airways

The configuration of the subcontinent, with high mountains to the north and ocean to south, India is mainly connected by international air routes from east and west. Air transport in India has registered a very high growth rate during past 30 years. The scheduled air services are run by Vayudoot, Indian Airlines and Air India. Indian Airlines provides air services within the country and to neighbouring countries. Air India provides services to the foreign countries. Major routes directly connect four cities, i.e. Delhi, Calcutta, Bombay and Madras, giving a linear or grid pattern to the routes. The difficulties of surface transport from northeastern states make airlines a vital assets to these states.

Considering the size of the country and need for rapid movement, airways play a vital role. It provides links between places of administrative, business, industrial and tourist importance. Out of about 85 civil aerodromes and 23 civil enclaves maintained by the

Ministry of Civil Aviation, the Indian Airlines touches about 57 points. The another domestic airline is Vayudoot which was first incorporated as a private limited company in 1981, and two years later became a public limited company. It was meant for the northeastern states initially, but later expanded to cover other places. Now it touches 50 points. Both the domestic airlines have several common points, such as Guwahati, Shillong, and Silchar.

Air India is basically meant for the international traffic but does have domestic service as well. It connects major Indian cities with about 50 cities of the world. Bombay, Calcutta, Delhi, Goa, Madras, Hyderabad, Trivandrum are included in the routes of this airline. It has got modern wide body aircrafts and is comparable with any other airlines of the world. Indian Airlines connects the important cities in the neighbouring countries. These cities are Kabul, Lahore, Karachi, Kathmandu, Dhaka, Chittagong, Bangkok, Singapore, Colombo and Male. There are several other foreign airlines which touch Indian airports, particularly Bombay and Delhi.

TABLE 20.3: Airways

Items	1971	1981	1986
Hours flown (000s)			
D	90	93	122
I	40	58	70
Kilometres flown (Mn.)			
D	33.4	44.4	63.1
I	25.2	43.7	46.3
Passenger carried (000s)			
D	2,056	5,556	9,180
I	491	1,944	2,316
Passenger km (Mn.)			
D	1,578	4,458	7,463
I	2,031	7,580	7,962
Passenger load factor (%)			
D	69.1	69.5	74.7
I	48.7	66.5	68.5
Freight tonne km (Mn.)			
D	14.1	59.8	91.0
I	97.1	367.9	447.4
Total tonne km. (Mn.)			
D	160	438	772
I	290	1,078	1,197
Revenue load factor (%)			
D	72.0	61.0	70.1
I	50.8	63.6	62.3

Source: **Statistical Outline of India, 1988-89.**
Note: **D - Domestic, I - International.**

The details of the domestic and international airlines are shown in Table 20.3. There has been drastic change in last 15 years (1971-86). Number of passengers of both types have increased many folds. The location and size of the country will still be able to sustain airlines with bigger fleet of aircrafts.

Means of Communication

The expanding economy of the country requires efficient means of communication from one part of the country to another, or within the same city. Indian Posts and Telegraphs Department plays a vital role in this regard. Apart from letters, the other means of communications are telegraph, telephone, telex and FAX. By the end of 1985-86 period, there were about 144,400 post offices in the country, out of which about 9 per cent were in urban areas. Telegraph offices were 36,000 in the same years. Though the number of telephone exchanges increased reasonably, the increase of telephone connections was significant. The telex exchanges and connections multiplied during this period. Apart from the post offices, other means of communications are urban based. Government through different programmes has been trying to connect each village by telephone facility (Table 20.4).

TABLE 20.4: Means of communication

Means of communication	1973-74	1980-81	1985-86	1986-87
Post offices (000)	116.8	139.2	114.4	n.a.
Urban (000)	11.5	14.5	15.6	n.a.
Rural (000)	105.3	124.7	128.8	n.a.
Telegraph offices (000)	17.1	31.5	36.0	36.6
Telephone connection (000)	1,637	2,785	4,044	4,409
Telephone exchanges (000)	20.3	29.4	37.7	38.4
Telex Exchanges (No.)	49	147	209	220
Telex lines (000)	8.5	19.3	30.5	34.0

Source: Statistical Outline of India.

Radio broadcasting was introduced in 1927, with privately owned transmitters at Bombay and Calcutta. Government took over in 1930 and 1936 and this service was called All India Radio (AIR). Since 1957 it is known as 'Akashvani'. The radio stations expanded from 6 in 1947 to 96 at present. The programmes are broadcasted from 173 transmitters of which 137 are medium wave. The radio network covers about 95 per cent of population and 83 per cent of the area of the country. It has become effective medium to inform, educate the entertain the masses. All the cultural and linguistic regions are covered by this network. It also has commercial services. AIR correspondents are all over the country and even in abroad.

Another means of mass communication is television, known as 'Doordarshan'. Experimental service was started in 1959, and regular service in 1965. But the expansion has been unprecedented in 1984 and after when one transmitter a day was installed. It now has 340 transmitters and serves about 80 per cent of the population. Further, there as 18 centres where programmes are produced. They are located all over the country. In the Seventh Plan such programme centres were to be increased to 48 with a total number of 423 transmitters. With the development of satellite technology, programmes originating from one centre can be transmitted through satellite. Again in 1984, second channel of transmission was introduced in the four metropolitan cities, i.e. Bombay, Calcutta, Delhi and Madras. Television has been used for entertainment, information, and educational purposes. Open university programmes are linked with Doordarshan. The basic aim is to provide programmes at three levels: national, regional and local.

There are other medium of mass communication as well. The most important one is the newspapers. However they have to depend a lot on the electronic media or electronic transfer of news items.

Transport, communication and electronic media have become very important for a diversified country like India. They provide a technological basis for the unity of different cultures, societies, regions and population groups. Development in the related fields will certainly helps in circumventing the tendencies which promote regionalism to undesirable extent.

REFERENCES

Das, P.K. & B.N. Sinha (1985), Inter-city airway connectivity in India, *Geographical Review of India*, Vol. XLVII(3), pp. 25-9.

Kara, P.K., S.N. Tripathi & D.K. Singh (1982), A comparative analysis of major ports of India and their traffic characteristics, *NGJI*, Vol. 28 (1 & 2), pp. 60-8.

NATMO (1985), *Glimpses of India*, NATMO Occasional Monograph, 2, Calcutta.

NATMO (1989), *Tourist Atlat of India*, NATMO, Calcutta.

NATMO (1990), *National Atlas of India*, NATMO, Calcutta.

Raza, Moonis & Yash Agarwal (1986), *Transport Geography of India: Commodity Flows and the Regional Structure of the Indian Economy*, Concept Publishing Company, New Delhi.

Singh, D.N. (1977), Transportation geography in India : Survey of research, *NGJI*, Vol. 23 (1&2), pp. 95-114.

21

PLACES OF TOURIST, HISTORICAL AND CULTURAL IMPORTANCE

India is land of great beauty and diversity. From the Himalaya to Kanyakumari the climate varies and so the people, their customs and traditions. India offers fine scenic contrasts. The snowy landscape of the Himalaya in the north, dry desert in the west, green tropical plain in the east and the bushy coconut, palms and tropical foliage growing along the western coast form scenic facets of India.

India's potentiality in developing tourism is vast and varied. Its scenic grandeur in the Himalayas, skiing facilities in many of its high altitude stations, fine beaches on its sea coasts, architectural beauty of many of its ancient historical temples and monuments, national parks and wildlife sanctuaries are some of the great attractions for tourists. In collaboration with state governments, the Indian Tourism Development Corporation set up by the Government of India is at present engaged in providing many facilities to the tourists by developing infrastructures, such as building new hotels and improving transport links for easy movement and comfortable housing. It is not easy to describe the places of tourist, historical and cultural importance of India briefly. Hence, in this study India has been divided into five regions, i.e. north, south, east, west and central to get an overall picture of tourism in the country (Map 21.1).

Northern Region

Northern zone consisting of states and union territories like Jammu and Kashmir, Himachal Pradesh, Punjab, Haryana, Rajasthan, Chandigarh and Delhi attracts largest numbers of tourist from India and abroad. Jammu and Kashmir and Himachal Pradesh both with their mountains, lakes and rivers have become renowned far and wide for ideal scenic beauty. Winter sports are major attraction in Kufri in Himachal

MAP 21.1: Tourist Map of India

Pradesh and Gulmarg in Jammu and Kashmir. Other centres are Simla, Dalhousie, Chamba, Kulu, Manali, Kasauli in Himachal Pradesh and Srinagar, Sonmarg, Pahalgam, Jammu in Jammu and Kashmir. There are many Hindu temples like Amarnath, Vaishnodevi, Baijnath etc. These two states have various starting points for trekking into the interior of the Himalayas. Punjab and Haryana have many towns like Jalandhar, Ludhiana, Patiala etc. which are tourist centres while Amritsar, a religious and historical centre of Sikhs, is also centre of tourist importance. Rajasthan identified as the land of princes has various places of historical importance. It has palaces on hill tops and lakes which are famous for their architectural design and beauty, e.g. Amber Palace in Jaipur, Lake Palace in Udaipur, and Chittor fort in Chittaurgarh. Rajasthan has some places of religious importance for Hindu, Muslim, and Jain in Puskar, Ajmer and Mt. Abu. Rajasthan is the only state in India from where scenic beauty of great desert can be seen. Jaisalmer along with Jodhpur and Bikaner form desert triangle. Bharatpur and Dhaulpur are famous for bird sanctuary and wild life sanctuary. This state is also famous for fairs and festivals of which *Gangor, Teej, Urs Moin-ud-Din Chisti* at Ajmer and *Pushkar* Fair are important. Delhi is the capital of India. The city has a fascinating history and stimulating present. New Delhi is the centre of administration, education, art galleries, museums, cultural institutions of modern India while old Delhi with its narrow lanes and houses is an epitome of the whole history of India. There are various places of tourist, religious, historical importance to see in and around Delhi. Chandigarh, the city constructed by famous French architect M. Le Corbusier is a tourist attraction on its own. It is capital of two states and one union territory.

Southern Region

The southern zone comprises of four states of Andhra Pradesh, Karnataka, Tamil Nadu and Kerala and the union territory of Pondicherry. Southern India has rich treasure of Dravidian culture along with the Indian classical dance and Carnatic music, folk arts and crafts.

This region has coastal plain, lush green tropical mountain, beautiful back water, famous temples and beautiful and historical cities. Andhra Pradesh attracts tourists for its historical places like Hyderabad, Warangal, Golconda Fort, religious centres, i.e. Tirupathi,

Sri Sailam; modern industrial centres, i.e. Visakhapatnam and Vijayawada. Karnataka is famous for historical places, e.g. Mysore Bijapur, Srirangapatna, Mangalore; religious places like Sravanabelgola, Sankaracharya Muth, and Chamundi Hills. Exquisitely carved temples of Belur and Halebid, beautiful Brindavan Garden, Ranganthitto bird sanctuary, Bandipur wild life sanctuary, graceful Jogfalls on Sharavati river and picturesque surroundings of Croog district are also important tourist centres. Bangalore, the capital city, is known as the City of Gardens. This is a city of eternal spring with beautiful gardens, famous educational and industrial centres. Kerala with her abundance of spices, ivory carving, sandalwood, and scenic beauty attracted many tourists for a long time. Kerala has numerous cities of magnificent buildings of ancient and historical architecture, famous shrines, beautiful sea beaches and wild life sanctuary. Places of tourist interest are Trivandrum, Cochin, Alleppey, Cannanore, and Trichur. Like any other states in the southern region, Tamil Nadu also has various historical religious industrial places. Tamil Nadu is famous for beautiful temples in Thanjavur, Madurai, Mamallapuram, Kanchipuram, Tiruvannamalai, Tiruchchirappalli. Tamil Nadu also has hill stations Udagamandalam, Kodaikanal and Yercaued. Pondicherry, originally a French territory, is now famous for the Aurobinda Ashram. The southern India has some exclusive festival like Floating Festival (Madurai), Meenakshi Kalyanam (Madurai), Pooram (Trichur), Onam (Kerala), Dusserra (Mysore), Pongal (Tamil Nadu).

Eastern Region

Eastern zone is unique in extending right from the Himalaya in the north to the sea in the south. The zones covers a large number of states in the eastern and northeastern part of the country. They are Arunachal Pradesh, Assam, Bihar, Manipur, Meghalaya, Mizoram, Nagaland, Orissa, Sikkim, Tripura, and West Bengal. Though the states in the northeastern region have good potentiality as a tourist centre but they have not developed properly due to lack of proper communication, and the restriction for foreigners to enter in some places. This region has some hill stations of which Shillong, the capital of Meghalaya, is important. Other places of interest are Guwahati, Sibsagar (Assam), Kohima (Nagaland), Imphal (Manipur), Agartala (Tripura). This region has perhaps the richest fauna in comparison to any part in Asia. It has Kaziranga reserve forest, Mannas sanctuary, Soni-Rupa sanctuary, Pabha Sanctuary for the protection of animals like one horned rhinoceros, wild buffaloes etc. Other four states of the eastern region i.e. Bihar, Orissa, Sikkim and West Bengal with Calcutta as their nerve centre attracts a large number of tourists. Bihar has numerous places of historical, religious, cultural and industrial importance along with exclusive tourist spots. Patna, Nalanda, Rajgirh are the places of historical importance while Deoghar, Bodh Gaya, Gaya are the places of religious importance. On the other hand, Jamshedpur, Ranchi, Sindri are industrial cities and Ranchi along with Netarhat Simultola, Madhupur, Maithan, Tilaiya, Hazaribagh are important tourist centres. Orissa, like Bihar, has numerous tourist centres, i.e. Puri (religious), Bhubaneshwar (religious and historical), Konark (rock carve temple), Chilka lake (tourist spot), and Gopalpur-on-Sea (tourist spot). Sikkim the Himalayan state in the eastern region, has beautiful scenic beauty. West Bengal, with its hill stations (Darjeeling, Karsiyang, Kalimpang), sea beaches (Digha, Bakkhali), historical centres (Murshidabad, Malda, Bishnupur), industrial areas (Durgapur, Chittaranjan) attracts large number of tourists. Calcutta, the capital of West Bengal is the main cultural and administrative centre of this region. Shanti Niketan, the work place of the poet Rabindranath Tagore and Viswa Bharti University, is an important cultural centre. Another important tourist spot in West Bengal is Sunderbans, the abode of the Royal Bengal Tigers. West Bengal and Sikkim provide ample scope of trekking in the Himalayan region. Like other parts of the country the eastern and northeastern regions have various fairs and festivals of their own, e.g. *Bihu* (Assam), *Saraswati Puja* (West Bengal), *Gangasagar Mela* (West Bengal), Car Festival (Orissa), *Raksha Bandhan, Durgapuja* and *Dussehra, Diwali*, Sonepur Cattle Fair, and *Karthika* festival.

Western Region

Western region comprises states of Maharashtra, Gujarat and Goa. Like eastern region, western region also has Bombay as its nerve centre. Bombay is the commercial capital of India and it is also a big tourist centre in India. The western region is very prosperous in industrial activities so it has large number of industrial towns. Pune is an important tourist centre. It is a major industrial city and historical centre, Aurangabad, Ahmednagar, Nagpur, Nasik, Kolhapur and Sholapur are historical, religious and cultural centres. The rock-cut temples of Ajanta and Ellora have become world famous now.

Mahabaleswar, Panchagani, Khandla, Lonavala, Matheran are hill stations in Maharashtra. Gujarat presents a unique combination of past and present with its most modern textile mills and chemical plants in Ahmedabad, Vadodara, Surat and Valsad. It is the one of the richest state in India. On the other hand, people of the Kachchh region in western Gujarat still practise nomadic culture. Porbandar, in Gujarat, is the birth place of Mahatma Gandhi. Present Gujarat consists of various princely states of British India. Hence, most of the towns like Vadodara, Rajkot, Jamnagar, Bhavnagar and Junagadh, have some historical and cultural background. Somnath and Dwaraka are important places of Hindu pilgrimage and Palitana is the sacred hill of Jains. Navsari is a celebrated place for Parsee Zoroastrians. Gir Forest is the sanctuary for lions, only of its kind in India. Goa is a picturesque land full of scenic charm and abounding places of tourist interest. It comparatively small size is totally belied by its many attractions such as historic forts, elaborately ornate churches and temples its evergreen hills and blue mountains, winding rivers, creeks and particularly palm-fringed pearl white beaches which leave any visitor to this place spell bound. Goa is the only place in India where tourists can witness carnival a colourful festival of Christians. Other special festivals of western region are Jamshed Navroz (Parsee), Mahavir Jayanti (Jain), Ganesh Chaturth and Navaratri.

Central Region

Central zone is the mid land of India, spread over two states — Uttar Pradesh, largest in population and Madhya Pradesh, largest in area. This zone include high mountains, vast plains and plateau. Uttar Pradesh with its lofty maintain, vast rivers and great plain is the most charming state of the country. It presents a grand mixture of Hindu-Islamic culture in its plain lands and simple tribal culture of the mountains. The whole Himalayan region of Uttar Pradesh is the most sacred religious centre for Hindus. According to Hindu mythology, this is the place of the God. This area is also famous for trekking and mountaineering. Haridwar, Rishikesh, Kedarnath, Badarinath, Gangotri, Gomukh, Yamnotri are religious places. Mussoorie, Nainital, Ranikhet and Almora are important hill stations. Western part of Uttar Pradesh presents many towns of Islamic history and culture like Lucknow, Agra and Aligarh. These cities are famous for their architectural structures, a delicate exquisite artistry, superb craftsmanship and art. Kanpur is an industrial town. Allahabad, Varanasi, Mathura, Vrindaban are the sacred cities of Hindu religion.

Sarnath is Buddhist religious centre. Madhya Pradesh also presents two elements mixed together — the Indo-Aryan and the aboriginal. This state has got a glorious history which can be traced in Bhopal, Gwalior, Ujjain, Mandu and Indore. The stone carving and architecture at the temples of Khajuraho shows medieval Indian art. Sanchi is a Buddhist pilgrimage centre. Jabalpur has one of the most beautiful scenic spots in India with marble rocks in the Narmada river. Madhya Pradesh has big industrial cities like Bhilai and Korba. Mandla and Bastar are places of tribal culture.

Islands

The India's two groups of islands, Andaman and Nicobar Islands in the Bay of Bengal and the Lakshadweep in the Arabian sea have picturesque landscape. Andaman Islands were a penal settlements from 1858 to October 1949, and except the aboriginals, almost the entire population consisted of convicts and of people who were in these islands in connection with these convicts. Several families from outside were brought to Andaman after Independence under the Pilot Scheme of Rehabilitation. Andaman and Nicobar Islands with its beautiful beaches, thick forest and blue sea offer very attractive prospect as tourist resorts. But these islands must have proper communication within themselves and with the main land. Government of India has set up Islands Development Authority to look after the overall development of all islands, and also to develop tourism to boost the economy.

Tourist Circuits

Within these five regions all the places mentioned earlier are not visited by tourists. It is not always possible to do so due to financial, transportational and time constrains. Hence, tourists are forced to select a few places thereby forming almost tourist circuits. These circuit may be intra-regional or inter-regional. For example Varanasi, Allahabad, Lucknow and Vindhyachal are in Central region, but the most popular circuit of Delhi, Agra and Fatehpur Sikri are in two region. Further, there are foreign tourists who travel by air and their movements are entirely different from the domestic tourists. Their preference would be Bombay, Delhi, Agra, Jaipur and Srinagar, or Bombay, Bangalore, Mysore and Madras. Then there are foreign pilgrims as well. The Hindus, mostly of Indian origin, would like to visit Mathura, Vrindavan, Varanasi, and Ayodhya. The Buddhists will like to see Sarnath, Patna and Bodh-Gaya. The Muslim tourist would pay visit to Jama Masjid in Delhi or Ajmer.

Foreign Tourists

It is relatively easy to keep record for foreign tourists as they are registered at the entry points, like international airports. Such tourists are also source of foreign exchange earnings. During 1971-81, their number has increased from 301 thousand to 1,163.8 thousand (3.8 times). These tourists are from all over the world (Table 21.1). The largest number is from United Kingdom (166,600) followed by the United State of America (134,900), Persian Gulf (77,000), Sri Lanka (74,400) and F.R. Germany (70,700). The most significant increase is in the number of tourists from U.S.S.R. (8 times), in past 16 years. The foreign exchange earning in 1987 was Rs. 1,890 crores.

TABLE 21.1: Foreign tourists visiting India

(in '000)

Nationality	1971	1980	1987
Afghanistan	6.2	10.8	11.8
Australia	10.9	22.6	32.4
Austria	1.8	7.2	8.0
Canada	7.0	23.8	37.7
France	16.8	58.7	64.4
Germany	17.9	54.7	70.7
Iran	2.8	20.9	23.6
Italy	6.0	29.0	41.2
Japan	11.6	30.6	46.2
Kenya	6.6	6.6	9.1
Malaysia	17.8	26.4	28.5
Netherlands	3.7	11.2	18.8
Persian Gulf	5.6	40.1	77.0
Saudi Arabia	0.9	23.5	24.5
Singapore	7.6	16.6	26.4
Spain	1.1	11.4	16.5
Sri Lanka	16.6	68.4	74.4
Sweden	2.8	7.1	11.2
Switzerland	5.4	13.3	27.8
Thailand	3.4	10.9	9.6
U.K.	40.4	102.5	166.6
U.S.A.	55.0	78.6	134.9
U.S.S.R.	3.5	11.0	28.0
Total (Incl. others)	301.0	800.2	1,163.8
Foreign Exchange earning (Rs. Crores)	n.a	1,166	1,890

***Source*: Statistical Outline of India, 1988-89, T.C.S., Bombay.**

Tourist Infrastructure

The places of interest and tourist circuits are related to the infrastructure provided in these places. In a bigger centre or hill stations, starred hotels are available which are meant for foreign and rich domestic tourists. Over 80 per cent hotels are starred, of which 30 per cent rooms are in five star hotels (Table 21.2). These hotels are well managed and have developed infrastructure. They are comparable to any hotel in western world. On the other hand, the infrastructure for poor domestic tourists is weak and in a way discouraging. Government is opening tourist lodges, *yatri niwas* etc. for middle class tourists. Several more, steps have to be taken in this regard, such as transportational facilities, rail reservation counters, eating places etc.

TABLE 21.2: Government approved hotels, 1987-88

Category	No. of Hotels	No. of rooms	Average occupancy ratio 1986 (%)
Five Star	56	10,303	69.8
Four Star	39	2,847	70.2
Three Star	107	5,924	65.5
Two Star	167	5,979	62.8
One Star	67	1,951	69.3
Unclassified	98	7,570	57.1
Total	534	34,574	67.7

Source: Statistical Outline of India, 1988-89, T.C.S., Bombay

REFERENCES

NATMO (1989), *Tourist Atlas of India,* National Atlas & Thematic Mapping Organisation, Calcutta, 2nd Ed.

NATMO (1985), *Glimpses of India,* National Atlas & Thematic Mapping Organisation, Calcutta.

Ramanaiah, J. (1989), *Temples of South India: A study of Hindu, Jain and Buddhist Monuments of the Deccan,* Concept Publishing Company, New Delhi.

Kaur, J. (1985), *Himalayan Pilgrimages and the New Tourism,* Himalayan Books, New Delhi.

Jha, M. (ed.) (1985), *Dimensions of Pilgrimage,* Inter-India Publication, New Delhi.

22

ECONOMY

During the post-Independence period of over four decade, the Indian economy has undergone a sea of change however with lot of internal and external pressures. Most of the factors had negative connotations which came in way of desired rapid transformation from the colonial to a self-sufficient economy. The progress has more so ever been alround. This country's development has surpassed the development of the ex-colonies under the Crown. Though the economy is primarily based on agriculture, it has been diversifying in all possible directions. In real terms, the growth of non-agricultural sector has been satisfactory. Wars, droughts, floods and the like name come in the way of economic development. However, targets of foodgrain or industrial output have generally been met.

Agricultural Sector

Indian economy is agricultural based. About 60 per cent of workers are engaged in this sector. The number of cultivators, who own land, is reducing, while the agricultural labourers are increasing (31.5 to 55.5 million in 1961-81 period). Its contribution in the net national product at current prices has reduced from 41.50 to 30.80 per cent during the decade ending in 1985-86, (Table 22.1). But, in actual terms, there has been an increase of 2-3 times. Obviously, the relative contribution is reducing continuously which is basically due to the comparatively faster growth of the other sectors of the economy, such as manufacturing, transport and communication etc. Further, decline in the agricultural output is likely to affect the net domestic product or national income. The target of the growth rate of the national economy cannot be met with poor performance of agriculture. This sector of economy is based on the rainfall during the monsoon period. The deficiency of rainfall during this period is certainly going to influence the economy of the country. For example, for the third year in

TABLE 22.1: Net national product at current prices

(per cent of total net domestic product)

Industry	1975-76	1977-78	1978-79	1979-80	1980-81	1981-82	1982-83	1983-84	1984-85	1985-86
1. Agriculture	41.50	40.23	38.02	35.72	37.22	35.06	33.04	35.41	32.67	30.80
2. Forestry/logging	1.00	0.94	1.05	1.10	0.98	1.05	1.02	0.83	0.87	0.80
3. Fishing	0.84	0.76	0.85	0.81	0.80	0.77	0.76	0.75	0.83	0.84
4. Mining & Quarrying	1.22	1.22	1.23	1.46	1.46	2.29	2.78	2.66	2.77	2.73
5. Manufacturing, Construction, Elec. & Water supply	21.08	22.30	23.22	23.87	22.75	22.55	23.29	22.67	23.11	23.91
6. Transport, Communication & Trade	19.05	19.11	19.48	20.50	20.40	21.66	21.67	21.64	21.49	21.73
7. Finance & Real Estate	5.34	5.74	5.97	6.13	5.87	6.02	6.08	5.74	6.28	6.63
8. Communities & personnel service	9.97	9.89	10.18	10.41	10.52	10.60	11.36	11.30	11.98	12.48
9. Total net domestic product at factors cost	62.557	75.939	81.477	88.543	1,05,445	120.093	1,34,488	159,256	175,489	197.078
10. Net factor income from abroad	(-)0.41	(-)0.31	(-)0.19	0.17	0.28	(-)0.01	(-)0.51	(-)0.62	(-)0.84	(-)0.75
11. Total net national product	99.59	99.69	99.81	100.17	100.28	99.99	94.49	99.38	99.166	99.25
12. Per capita	1,026.4	1,194.1	1,253.0	1,336.1	1,557.3	1,743.0	1,887.3	2,186.0	2,258.8	2,595.6
Increase (%) in Net Domestic Product (Base 1975-76).	-	21.39	30.24	41.57	68.56	43.38	114.98	154.58	180.53	215.20

Source: Indian Agriculture in Brief.

continuation, rainfall during 1986-87 was unfavourable as well. Out of 35 meteorological regions, rainfall during this year was scanty or deficit in 14 regions as against 9 in 1985-86 and 8 in the preceding year. Due to this reason, the foodgrain production during 1986-87 is about 10-11 million tonnes below the target of 160 million tonnes.

Efforts have been made to develop infrastructure for higher agricultural production despite of floods and droughts. About 28 per cent of the net area shown is now irrigated. The irrigated area has only increased 91 per cent since 1950-51. However, several river valleys, canals (major and minor), tanks and tubewells have been initiated. Over 30 per cent of the area under rice and wheat is irrigated. Marketing and procuring co-operatives have been encouraged. Agricultural credits have been advanced through such co-operatives and banks (Rs. 568 crores in 1983-84). Agro-economic research centres in the major states have several development programmes which have been initiated for long and short term targets. Measures to check soil erosion, consolidation of operational holdings, river valley projects, use of tractors and fertilisers are some of the attempts to support higher production in the agricultural sector. Owing to such an infrastructure, the area under principal crops remained almost same during last decade or so, but the production has increased substantially.

Apart from forestry and fishing, other sectors have recorded increasing share in the net national product (Table 22.1). In actual value, mining and quarrying (7 times), manufacturing etc. (3.6), transport and communication (3.6), finance and real estate (3.9) and community and personnel service (3.9) have multiplied several times. There is a clear trend of the diversification of the economy which is a healthy development.

Energy Sector

Different features of the economy have indicated development all over. Due to timely planning for electricity, its generation has increased from 5.3 billion kWh in 1950-51 to 201.8 billion kWh in 1987-88. Though, such development has been beneficial to the establishment of new industries at large, there has been a scarcity of power. Infrastructure is ready to consume more electricity. This is not only due to multiplying of industries but also for the increasing demand of the use of electricity in domestic sector due to reduction of conventional sources of energy and advent of modern appliances. Per capita domestic use of electricity has increased from 1.6 kwh in 1950-51 to 22.7 kWh in 1985-86 per year.

The other sources of energy which have made impact in the Indian economy are petroleum and natural gas. The oil map of the country has changed since 1970-71. The production of crude oil was only 6.8 million tonnes then which became 30.5 million tonnes in 1986-87. Petrol has been found on-and off-shore, and in several areas there are positive indications. Due to the domestic production of crude oil, imports have been restricted. The variation has been less. In 1970-71, only 11.7 million tonnes of oil was imported which rose to 15.6 million tonnes in 1986-87. On the other hand, almost whole of petroleum products required are produced in the country (import 1.4%). However, there has been a sharp increase in the consumption pattern. During the period 1970-71 to 1985-86, consumption has been increased about 2-5 times. The petroleum products are light distillates (naphtha, motor gasoline), middle distillates (kerosene, high speed diesel oil), and heavy end (furnace oil). Consumption has increased substantial in high speed diesel oil — 4.2 times in 15 years. The consumption of petroleum product is due to multiplication of motor vehicles of all sorts from 1.87 million to 10.2 million in last 15 years. Two wheelers, which suit Indian urban and rural conditions, have multiplied over ten times during this period. It has been considered as a symbol of expanding middle class affluence. The consumption of petroleum products in agriculture is negligible.

Natural gas is being produced on-shore and off-shore. The off-shore production became significant from 1980 onwards. In total, gas from both the sources, increased from 1,445 million m³ in 1970-71 to 9,853 million m³ in 1986-87. Due to the re-injection and flaring, about half of the gas produced is lost. However, urban domestic consumption is ever-increasing. About 81 per cent of L.P.G. (cooking gas) is for this purpose.

Mineral Resources

Another major source of the natural wealth is from minerals. India is a major produce of coal, iron bauxite and manganese. Recoverable reserves likely at the end of 1994 for non-cooking coal is 30,189 million tonnes, cooking coal is 6,006 million tonnes, iron ore of all grades is 8,815 million tonnes, bauxit in 2,106 million tonnes, and manganese ore of all grades is 54 million tonnes. The balance life after this year will be 116, 93, 32 – 182, 301, and 15-31 years respectively.

Further, there is some relationship between the reserves and production. The mineral having limited life span, the production has gone down. Only 71 per cent of manganese ore and 51 per cent of gold of 1971 is now being produced. The production of other minerals have increased however at varied proportion.

Industrial Sector

Based on agricultural products, minerals and energy, industry is bound to develop. But, as mentioned earlier, the inherited colonial infrastructure, capital and to certain extent know-how have come in this way of desired development. The number of factories increased steadily from 40.5 thousand in 1960 to 96.9 thousand in 1984-85. The peak year was 1981-82 when the number increased to 105 thousand. During this period of about two and half decade employment in factories increased from 3,644 to 7,870 thousand. The largest number of such units have invested a capital less than Rs. 10 lakhs (35.5% factories) and having employees less than 50 persons (77.5%). There has been a tendency of concentration of factories in certain states, e.g. Maharashtra (16.1%), and West Bengal (11.4%). But, the output of the latter state is only 8.1 per cent of gross national output from factories. The largest number of factories are engaged in food products, cotton textiles and electricity. Over 92 per cent of factories are wholly private owned. In addition, there are small-scale, and khadi and village industries. In the former case, the number of units in 1986-87 were 1,457 thousand and the employment is about 10 million. West Bengal, Madhya Pradesh and Uttar Pradesh have concentration of these units. On the other hand, the khadi and village industries provide full time and part-time employment. Out of about 4,070 thousand persons engaged in this sector, about 65 per cent are part-time. However, the employment in both have doubled in last 15 years. The industries in general have diversified since 1970. Some industries have expanded significantly, such as TV, tractors, dry cells, power transformers, polyester fibre, electronics, computers and commercial vehicles.

Some Indicators of Economic Development

There are other sectors of economy where the development is apparent. These developments are reflected by the increase in total road length (3.38 times since 1960-61), number of trucks and buses on road (4.66 and 3.74 times respectively), electrification of rail lines (about 2 times since 1970-71), and number of passengers using rail lines (4.46 times). The list of such indicators highlighting the growing economy is endless (Table 22.2). The export of traditional items such as tea, coffee, cashew, kernels, spices, iron ore, cotton and jute is increasing. In some cases the exports has multiplied several times in past 15 years. Further, there are non-traditional items as well, such as fish and meat and their preparations, garments, leather goods, iron and steel, and engineering and chemical goods. In some cases the export has been phenomenal. Nevertheless, there has been a continuous deficit in import-export equation, and this is increasing every year. In fact, in several items, the proportion of the Indian contribution to world export is reducing. This is due to restrictions, sanctions and protectionist tendencies in different countries.

Even with all round development in agriculture, energy, industry, mineral and other sector of the economy, there has been. regional variations in the economic development. Some states are well advanced, while in other states and union territories the development in true sense is yet to be realised. Further, the development processes have been concentrating in and around the larger urban centres including mega-cities. They continue to attract development despite of governments' efforts to diversify the economy.

TABLE 22.2: Economic development — some indicators

Items	1950-51	1970-71	1986-87/ 1987-88
1. Steel ingot (million tonnes)	1.5	6.1	12.4
2. Cotton Cloth (Billion Metres)	4.2	7.8	10.8
3. Surface Roads ('000 km)	157.0	398.0	N.A.
4. Shipping tonnage ('000 GRT)	391.0	2,397.0	5,675.0
5. Railway freight (Million tonnes)	93.0	197.0	318.0
6. Exports (Rs. Crores)	601.0	1,535.0	15,719.0
7. Imports (Rs. Crores)	650.0	1,634.0	22,343.0
8. Export on % of import	92.6	9.4	70.4
9. Wholesale Price Index (1970-71 = 100)	48.0	100.0	405.0
10. School students (Million)	22.0	106.0	117.0
11. Hospital beds ('000)	113.0	599.0	645.0

Source: Various official data.

The economic indicators will clearly reveal that in the economically advanced states will be having (a) higher percentage of urban population, (b) more irrigated land, (c) high consumption of fertiliser per ha, (d) larger number of workers in organised sector, (e) high consumption of electricity, (f) more villages connected by all roads and electricity, and (g) better social facilities. However, the per capita income of the states and union territories is not only due to the stage of economic development but also owing to the availability of central (federal) funds. The high income in Delhi (Rs. 5,315 in 1985-86) and Pondicherry (Rs. 3,777) is due to this reason. On the other hand, Punjab (Rs. 4416), Haryana (Rs. 3,669) have better agricultural base (Map 22.1). Similarly, the higher income in the state of West Bengal (Rs. 2,813) and Maharashtra (Rs. 3,430) is due to its industrial and resource base (Table 22.3). Obviously, a single sector of the economy is not only responsible for such development. However, one sector of the economy tends to attract the other sectors as well. The surplus in one paves way for the development of others. Indian economy is considered to be at a take-off stage. Some sectors have performed well, however others are lagging behind. The central issue in the economic development of India is the spatial inequalities which has lead to several social, economic and political problems. Balanced economic development is considered as the key to solve several problems which the country is facing today.

TABLE 22.3: Net domestic product and per capita income (current prices), 1985-86

States/Union Territories	Net domestic product (Rs. Crores)	Per capita income (Rs.)
Andhra Pradesh	12,757	2,184
Assam	4,638	2,017
Bihar	11,929	1,548
Goa, Daman & Diu*	525	4,443
Gujarat	10,639	2,772
Haryana	5,379	3,664
Himachal Pradesh	1,187	2,592
Jammu & Kashmir	1,459	2,173
Karnataka	8,753	2,136
Kerala	6,322	2,287
Madhya Pradesh	11,468	1,988
Maharashtra	23,731	3,430
Manipur*	344	2,200
Orissa*	4,316	1,534
Punjab	8,180	4,416
Rajasthan	7,921	2,043
Tamil Nadu	12,220	2,353
Uttar Pradesh	24,260	1,988
West Bengal	16,810	2,813
Delhi	4,009	5,315
Pondicherry*	253	3,777

* Data relates to 1984-85.

Source: Statistical Outline of India, 1988-89.

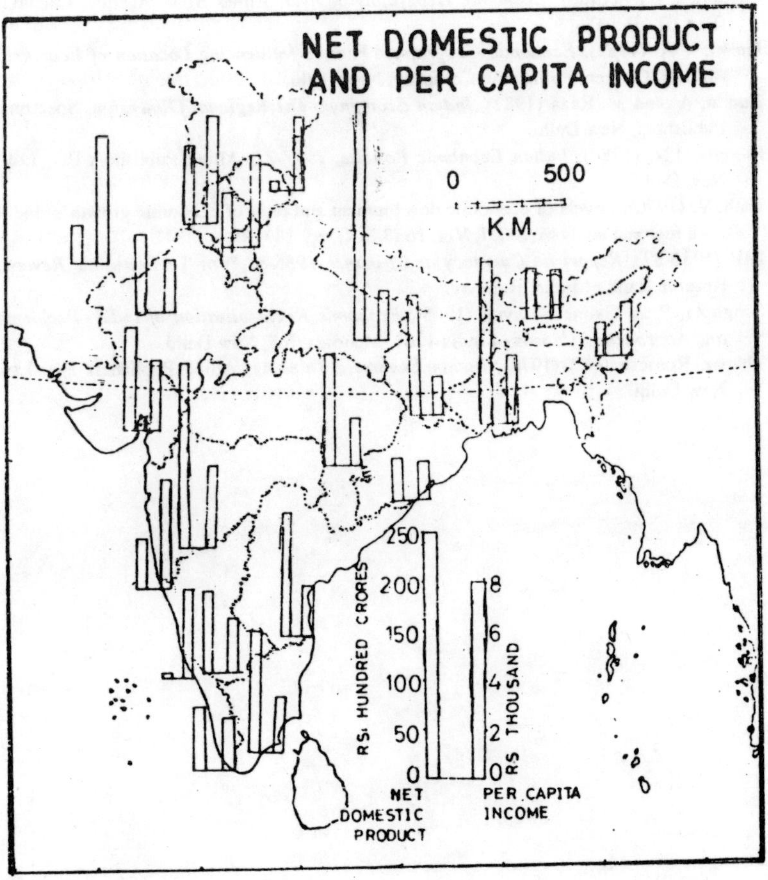

MAP 22.1: Per Capita Income

REFERENCES

Chadha, Rajesh (1989), *A Key Sector of Indian Economy: Systems View of Steel Industry*, Concept Publishing Company, New Delhi.

Chatterjee, S.P. (1986), *Economic Geography of Asia*, Allied Book Agency, Calcutta, (Part III - India).

Janaki, V.A. (1985), *Economic Geography: Factors Influencing Location of Economic Activity*, Concept Publishing Company, New Delhi.

Kundu, A. and M. Raza (1982), *Indian Economy : The Regional Dimension*, Spectrum Publishers, New Delhi.

Mongia, J.N. (1980), *Indian Economic Policies, 1947-71*, Allied Publishers Pvt. Ltd., New Delhi.

Nath, V. (1970), Levels of economic development and rates of economic growth in India — a regional analysis, *NGJI*, Vol. 16 (3 & 4), pp. 183-98.

RBI (1986-87), *Report on Currency and Finance 1986-87*, Vol. I : *Economic Review*, Reserve Bank of India, Bombay.

Sengupta, P. & Galina Sdasyuk (1968), *Economic Regionalisation of India: Problems and Approaches*, Census of India 1961, Monograph 8, New Delhi.

Thapar, Romesh (ed.) (1979), *Indian Economic Thinking*, Allied Publishers Pvt. Ltd. New Delhi.

23

INDIAN OCEAN

Historically and culturally, Indian Ocean has always been closely associated with India. Its location at the vertex of the vast ocean is unique. India became focal centre of transcontinental trade and commerce route cutting across the ocean. In early days, navigation routes connecting the ports in the littoral states were possible. Later, navigators were able to find direct routes as well, e.g. from West Indian coast to East African coast. As the controller of the Indian Ocean trade, India became a basis of empires built by the European powers. It must, however, be emphasised here that the trade, commerce and settlement among the people of the littoral states started much before the advent of the European powers on the scene.

Unlike other oceans, this Indian Ocean has never been a dividing factor among nations, states, regions and cultures. In fact, it has brought the diverse cultural groups of different parts of Asia, Africa and Europe together which was ultimately developed into Indian Ocean culture. It is the most humanised ocean of the world and no major war has been fought over it.

Extent of the Indian Ocean

The extent of this ocean has defined in different ways by different people. Vallaux puts its southern limits at 35° south while Schott includes the southern sea in it. The eastern limits of the Indian Ocean are also debated and there is yet no agreement whether Timor and Arajura seas lying between Indonesia and Australia be included in it. However, geographically the boundary between the Pacific and the Indian Oceans follow the line from Malay Peninsula through Sumatra, Java, Timor, Australia (Cape London Derry) and Tasmania and follows the meridians of 147°E to Antarctica. The 20°E Meridian is considered to be the dividing line between Atlantic and Indian Oceans (Map 23.1). It passes through the Cape of Good Hope in Republic of South Africa.

Indian Ocean

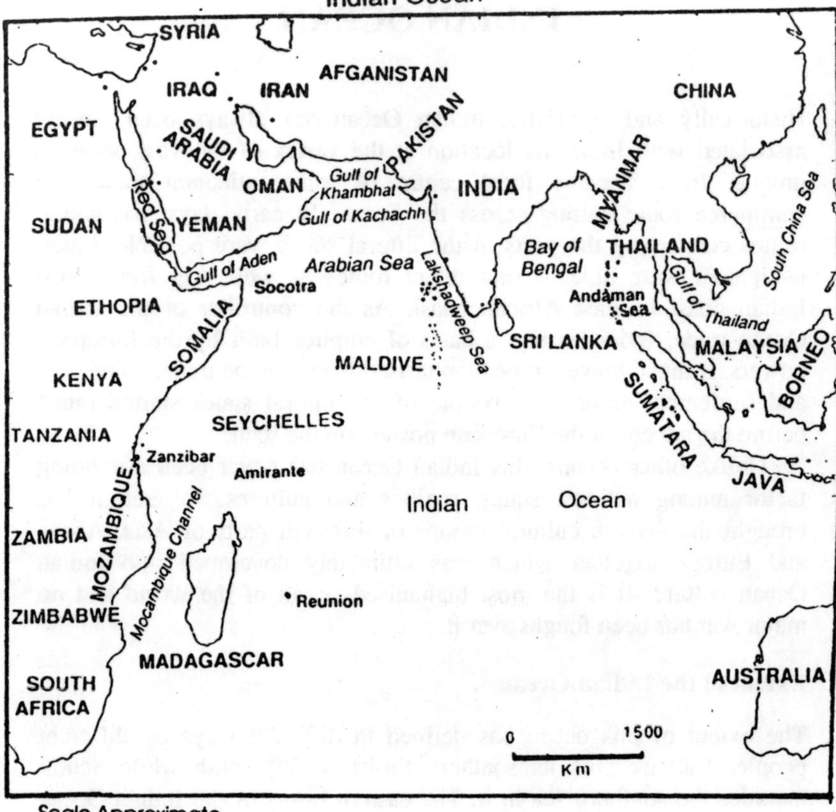

Scale Approximate

MAP 23.1: Indian Ocean

The Indian Ocean has limited outlets. Suez canal is an artificial link with the western world. Another possible link with Europe is by circumferencing whole of Africa. On the eastern side there are two outlets – one through the islands of Indonesia, and second is via south of Australia. Indian Ocean can be chocked anytime by controlling these outlets. With rich resources in this ocean and in the littoral states, such a possibility has added an additional dimension in the geopolitical strategy of the region. This is not only the factor for the continuing interest of the former colonial powers.

Indian Ocean consists of marginal seas touching the littoral states. They are Malagassi Sea, Red Sea, Gulf of Aden, Persian Gulf, Gulf of Oman, Arabian Sea, Laksadweep Sea, Palk Strait, Bay of Bengal, Andaman Sea, Singapore Strait, Suva Sea, Timor Sea, Arajura Sea, Gulf of Carpentaria, Torre Strait, Exmouth Gulf, Great Australian Bight, Spencer Gulf and Bass Strait. In comparison to other oceans, it has the largest number of marginal seas – touching warm water of the Arabian coast to the frozen landmass of Antarctica.

Littoral States

There are about 43 littoral states which including the Indian Ocean forms the Indian Ocean Realm. These countries are considered to be associated with this ocean, traditionally, culturally and economically. Out of 43 countries, 19 are in Africa, 23 in Asia and one in Oceania (Table 23.1). There are also land-locked countries, such as Zambia, Uganda, Zimbabwe, Nepal, Afghanistan, which are also included in this realm because their natural outlets are into this ocean. On the other hand, there are island states, such as Malagassi, Mauritius, Sri Lanka, or Australia, forming an essential part of this realm. In addition, there are islands which are parts of different countries. The examples are of Andaman and Lakshadweep islands of India; Sumatra, Java and several other islands of Malaysia; Zanzibar of Tanzania; or Socotra of Democratic Republic of Yemen. Further, there are islands which are under the control of former colonial powers. The islands of Seychelles, Amlrante, Aldabra, Chagos Archipelago belong to Britain, while Reunion, Comoro, Crozet or Kerguelen are under the French administration. These islands have played very important role in the development of trade and commerce in this realm. The southern portion of the Indian Ocean is formed by the lesser known continent of Antarctica.

TABLE 23.1: Countries in the Indian Ocean realm

Continent	No. of countries	Name of countries
Africa	19	Botswana Burundi, Egypt, Ethiopia, Kenya, Lestho, Malagassi, Malawi, Mauritius, Mozambique, Somalia, Sudan, Swaziland, Rwanda, Uganda, Tanzania, Zambia, South Africa, Zimbabwe
Asia	23	Afghanistan, Bahrain, Bhutan, Myanmar, N. Yemen, S. Yemen, India, Indonesia, Iran, Iraq, Kuwait, Malaysia, Maldives, Nepal, Oman, Pakistan, Qatar, Saudi Arabia, Singapore, Sri Lanka, Thailand, Bangladesh, United Arab Emirates.
Oceania	1	Australia.

Antarctica

It is believed that about 200 million years ago Antarctica was a part of a supercontinent which included South America, Africa, India and Australia. Over million of years the snow has been depositing over it which now has reached an average depth of two kilometres, thereby becoming the biggest reservoir of fresh water on the earth. It covers 98 per cent of the landmass. Antarctica is also the highest continent with an average elevation of about 1800m. Its 13.5 million square km landmass is considered to be the last frontier. During the International Geographical Year (1957-58), 12 countries conducted extensive expeditions on this region and established 50 stations. This has lead to the formulation of Antarctica Treaty which made the provision that this territory shall be used for peaceful purpose only and for scientific exchange without territorial claims. The original 1959 signatories were Argentina, Norway, South Africa, U.S.S.R., U.K., U.S.A., New Zealand, Australia, Belgium, Chile, France and Japan. Since then Brazil, India, Uruguay, China, Poland and F.R. of Germany have become full consultative parties. They are conducting expeditions and have research station on this landmass (Fig. 23.2). This treaty will be reviewed in 1991. The portion of Antarctica which touches the Indian Ocean is claimed by Norway and Australia. The future of this global common is unknown. However, rich mineral resources and petroleum reserves are likely in this continent or in off-shore areas. India has sent several expeditions teams there and have a permanent research station at Dakshini Gangotri (70°45′12"S and 11°38′13"E). Another station, Maitri, has been established 70 km south of the earlier station.

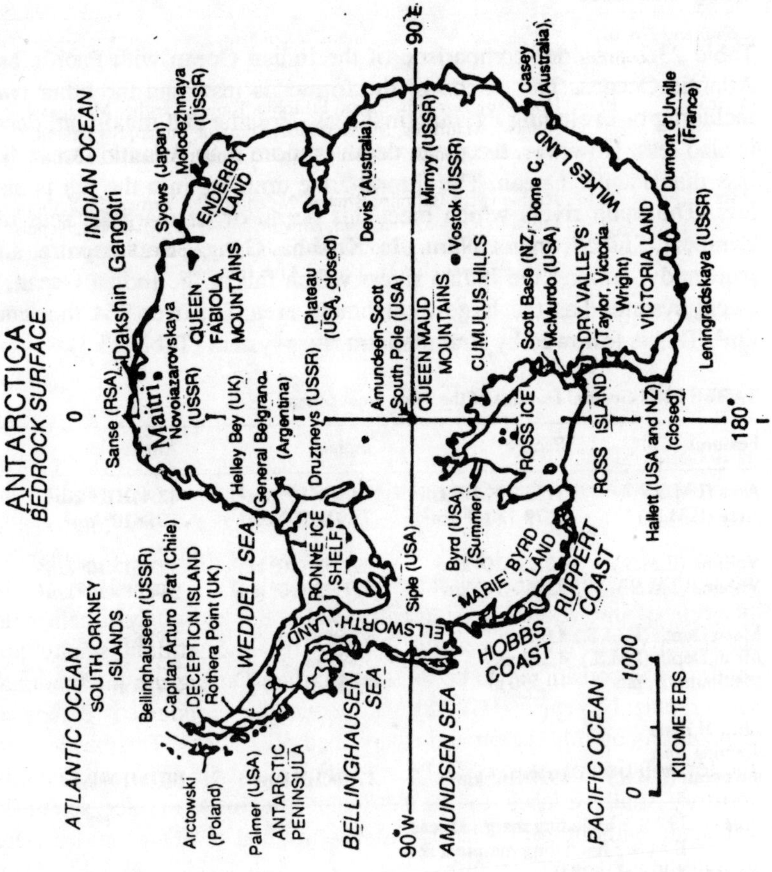

FIG. 23.2: Antarctica

Other Features

Table 23.2 provides comparison of the Indian Ocean with Pacific and Atlantic Oceans. The extend of the former is less than the other two, including or excluding the marginal seas. Volume and maximum depth is also less. However, the mean depth is more than Atlantic Ocean but less than Pacific Ocean. The littoral zone draining into the sea is also less. The main rivers which meet this ocean or its marginal seas are Zambezi, Shibeli, Indus, Narmada, Krishna, Ganga, Brahmaputra, and Irrawaddy. Among the Indian rivers which fall in the Indian Ocean, 6 rivers systems has the largest catchment area, i.e. over 964 thousand km². This is followed by Brahmaputra river system (Table 23.3).

TABLE 23.2: General features of the three great oceans

Features	Pacific	Indian	Atlantic
Area (I.M.S.)	1165.2×10^6 km²	74.92×10^6 km²	82.44×10^6 km²
Area (E.M.S.)	1179.7×10^6 km²	75.95×10^6 km²	106×10^6 km²
Volume (E.M.S.)	707.5×10^6 km²	291.03×10^6 km³	323.61×10^6 km³
Volume (I.M.S.)	723.7×10^6 km²	291.95×10^6 km³	354.68×10^6 km²
Mean Depth (E.M.S.)	4,028 m	3,963 m	3,926 m
Mean Depth (I.M.S.)	4,282 m	3,897 m	3,332 m
Maximum Depth	10,550 m	7,437 m	8,750 m
Littoral Zone draining into the ocean (I.M.S.)	19.4×10^6 km²	17.0×10^6 km²	50.0×10^6 km²

Note: I.M.S. : Including marginal seas
E.M.S. : Excluding marginal seas
Source: S.K. Pal (1984).

TABLE 23.3: Rivers and the Indian Ocean

River system	Catchment Area (km²)
1. Rivers falling into Arabian Sea (Excluding the Indus System)	4,85,862.40
2. Indus Basin in India	3,49,882.88
3. Rivers falling into Bay of Bengal Other than Ganga and Brahmaputra	11,96,311.00
4. Ganga System	9,64,654.08
5. Brahmaputra System	5,00,377.60
6. Rajputana	1,66,110.72

Source: A.N. Khosla, *Appraisal of Water Resources: Analysis and utilisation of Data*, UNSCCUR Proceedings.

Mapping of the Indian Ocean

Mapping of the Indian Ocean has been started much earlier. The role of Portuguese cartographers in this regard has been commendable. Cartography was a part of their activities related to hydrographic exploration, discovery and imperial colonisation. The first navigation charts to reach Africa and India is claim to have been prepared by Andrea Bianco (1448) and Graziano Benincasa (1465) of Portugal. There are several Portuguese contributions in mapping and charting the Indian Ocean or its parts or parts of India. However, their attention was more on Goa and Diu. Portuguese maps and charts have been copied by the cartographers and cosmographers of other European countries during the days of great discoveries.

With the supremacy of the British power in the Indian Ocean and in the important littoral states, the hydographic surveys were established. It became the part of the mandate of these institutions to prepare maps and charts of this part of the Ocean and the coast. It was basically meant for navigation. But, with the oceans becoming a new frontier to tap resources, particularly petroleum, the interest in mapping has been in relation to exploration of resources and strategic considerations. Atlas of this Ocean has been prepared by the Russians. A plate on this Ocean showing Ocean currents and seasonal wind direction has been included in the National Atlas of India. Further, in the same publication, there are maps on the Bay of Bengal and the Arabian Sea.

Earlier, only the ship observation have been the main source of data for mapping over oceans until recently with the advent of satellites and their capability to sense in several regions of the electromagnetic specturm, the situation has changed drastically. The satellites provide a synoptic view over vast regions and the data obtained is uniform and compatible to modern methods of computer data procurring. Several countries are having their own satellite programme and a number of them do provide data over the Indian Ocean. The Indian National Satellite, INSAT-IB which is geosynchronous and provides three hourly observations. The very high resolution Radiometer (VHRR) operates in the visible region (2.75 km resolution) and thermal intrayed regio (11 km resolution) and provides information on sea-surface temperature, cloud cover, cloud level winds.

There are polar orbiting satellites as well, such as NOAA-9. The on-board sensors are Advanced Very High Resolugion Radiometer (AVHRR). The data available can be used for mapping of sea-surface temperature, cloud cover, vertical temperature profile etc. Further, the earth resource satellites, such as LANDSAT, SPOT, IRS - 1A provide information in the optical region and can be used for mapping for sedimentation, ocean colour and the ocean pollution studies.

Though, the above sources are available to provide information for mapping, there is hardly any attempt to map different themes for the Indian Ocean. Pollution of Indian Ocean is an important issues. The wastes of thickly populated region in the littoral states, industries, increasing maritime activities, indiscriminate fishing, oil spilling etc. have been endangering the Indian Ocean marine environment. In addition, industrialised nations have been dumping nuclear wastes here. The possibility of the rise of the ocean huels due to Green-house effect has been a cause of concern all over.

The nations around the Indian Oceans are still developing, with additional population and industries, this ocean is likely to be more and more polluted. The most humanised oceans may turned to be most polluted and a place of all sorts of conflicts. Steps must be taken right now to circumvent any undesirable situation.

REFERENCES

Aurnachalam, B. (1987), The haven-finding art in Indian navigational traditions and cartography, in Satish Chandra (ed.), *The Indian Ocean Explorations in History, Commerce and Politics*, Sage Publications, New Delhi, Newbury Park, London, pp. 191-221.

Chaudhuri, K.N. (1985), *Trade and Civilisation in the Indian Ocean*, Cambridge University Press.

Correia, A fonso, John (1984), Luso-Indian Cartography, Paper presented in the International Conference on Indian Ocean Studies, Proceedings, Perth, December 1984 (mimio). Also published in *Indica*, Vol. 22, No. 2, 1988, pp. 123-33.

IUCNNR (1985), Antarctica, Key Environment series, Pergamon Press, Oxford in Collaboration with the International Union for the Conservation of Nature and Natural Resources.

Nag, P. (1987), The Indian Ocean, India and Africa: Historical and geographical Perspectives, in Satish Chandra (ed), *The Indian Ocean: Explorations in History, Commerce in Politics*, Sage Publications, New Delhi, pp.151-23.

Nag, P. (1979), Asian Settlement in Zambia, *Africa Quarterly*, Indian Centre for Africa, Vol. XVIII, No. 4, pp.55-77.

Nanda, J.N. (1989), *Development of the Resources of the Sea: India*, Concept Publishing Co., New Delhi.

Pal, S.K. (1984), Cartography of the Indian Ocean region, Paper presented in the International Conference on Indian Ocean Studies, Proceeding F. Perth, December 1984, (mimio).

Panikkar, K.M. (1945), *India and the Indian Ocean*, Bombay.

Raza, Moonis (1985), The Indian Ocean — A Geographical Perspected, paper presented in the International Seminar on the Indian Ocean, New Delhi, Feb., 1985.

Saxena, P.B. and Satya Prakash (1977), The North Indian ocean: A Study of the impact of the anemo-hydrography on the early marine sailing routes, *NGJI* Vol. 23 (3 & 4) pp.125-32.

Singh, T.D. (1988), *Hind Mahasagar Avam Parimandalya Rashtra — Aik Bhogolika Adhyanyan*, Tara Book Agency, Varanasi (in Hindi)

Stone, Jeffrey C. (1985), *Africa and the Sea*, University of Aberdeen, Aberdeen.

24

PROBLEMS AND PROSPECTS

The most serious problem which India is facing today is its teeming population. Developments of all sorts are engulfed by the ever-increasing numbers. Unemployment is becoming widespread, and the urban centres are the focus of migration of rural unemployed persons. The cities are becoming unmanageable, suffering from pressure of population and from pollution. Both are increasing continuously. Their growth is retarded due to urban famine and power crisis. On the rural scene, the population problem is no less acute. The percentage of cultivators, i.e. persons owning land, is decreasing continuously, while the percentage of agriculture labours, i.e. landless workers, is also decreasing. There has been an increase in the percentage of 'other workers', such as social, and personal services, transport, construction etc.

Population-related Issues

The second most populated nation is increasing its population at the rate 2.25 per cent per annum (1971-81). However, some states of this union have recorded decline in the growth rate. But, the populous states, like Uttar Pradesh and Bihar, have higher rates in comparison to the earlier decades. The percentage of rural population has declined from 80.1 per cent in 1971 to 76.7 per cent in 1981. The sex ratio continues to be adverse to females. The percentage of population in the age group 0-14 has declined in 1981 in comparison to the earlier census. The percentage of workers have increased marginally during 1971-81. The female participation has increased reasonably (14.15 to 19.76%) in general, but their share as agricultural labourers or service workers has decreased.

The population continues to increase due to fall in crude death rate. This has neutralised the benefits of the decreasing birth rate. This is due to better health facilities and awareness. The largest number of deaths is due to diseases of the circulatory system (20.7%) in urban areas, and senility (22.4%) and coughs-disorders or respiratory system (19.8%) in rural areas. There is a significant variations in the infant mortality rate in rural (105) and urban (62) areas. The trends clearly indicate that the fertility has decreased and the life expectancy has increased. It is estimated that population would be 83.72 crores in 1991 and 98.61 crores in 2001 A.D. Considering the increase in the age at marriage reduction in incidence of widowhood and likely achievements in family planning, the crude birth rate is expected to decline from 34.7 in 1980 to 23.7 by 1996-2001. This would imply a decline in the growth rate of population from about 2.2 per cent in 1981 to 1.5 per cent by 1996-2001. But the so far success claimed by the family planning has been challenged. Due to the growth of population, a larger number will be included in the work force for which sources of employment will have to be planned well in advance. The situation in the urban areas will become more acute due to doubling of population in 2001 A.D. (Table 24.1).

TABLE 24.1: All India percentage distribution and increased of population in age-groups, 1981-2001

Age group	1981	1986	1991	1996	2001	1981-2001
0-4	14.17	13.54	12.92	11.93	10.74	9.73
5-9	13.45	12.14	11.73	11.37	10.69	14.99
10-14	12.08	12.01	10.90	10.64	10.44	25.03
15-19	10.14	10.82	10.81	9.90	9.78	39.59
20-24	8.55	9.05	9.71	9.80	9.09	53.84
25-29	7.44	7.62	8.11	8.79	8.99	74.83
30-34	6.47	6.63	6.82	7.34	8.06	80.15
35-39	5.71	5.75	5.92	6.16	6.71	70.03
40-44	5.04	5.04	5.10	5.31	5.60	60.87
45-49	4.35	4.40	4.43	4.55	4.80	59.71
50-54	3.55	3.75	3.82	3.90	4.06	65.53
55-59	2.85	2.98	3.18	3.29	3.41	73.32
60-64	2.26	2.29	2.43	2.64	2.78	78.32
65-69	1.68	1.70	1.76	1.90	2.11	82.19
70 +	2.27	2.28	2.35	2.49	2.74	74.56
All ages	100.00	100.00	100.00	100.00	100.00	44.71

Source: Census of India (1981), Population Projections for India, 1981-2001, Paper 1 of 1984.

The urban population or the population residing in about four thousand urban centres of different sizes is about 23 per cent which is low even by developing countries standards. But, its growth has been significant during the past census decade (46.8%). The urban population, excluding Assam, is about 16 crores of which about 60 per cent live in the cities having more than one lakh population. On the other hand, the population in the towns with a 20,000 persons or less is decreasing (14% in 1981). However, larger number of towns are in this group (20% in 1981). This due to migration to the mega-cities. The small towns have been unsuccessful in checking this trend owing to less employment opportunities there. In cities like Ahmedabad, Bombay, Calcutta, Delhi, Kanpur and Pune, 60 to 76 per cent of the migrants came for employment are from rural areas. Migrants form 33 to 50 per cent of the population in these cities. The excessive migration have created pressure on urban amenities and over-crowding. About half of the households in these mega-cities live in single rooms. The basic facilities like tap water, electricity, toilet are becoming scarce.

Environment-related Issues

Due to the increase in population the environmental problems are not only confined to the urban areas. There has been population pressure on agricultural area as well. India has already one of the highest percentage of land under agriculture and there is very limited scope to bring more land under this use. However, there has been an attempt to put the wasteland under more productive uses. According to the Planning Commission's estimation in 1981, about 57.56 per cent of the area of the country are subjected to the one or other environmental problem. Everyday additional land, and population is being exposed to new environmental risks. Thereby the areas falling in the multiple risk zones have been enlarging. One of the major environmental problem in rural areas is the soil erosion. About half of the country's areas is affected by this problem. Of the total cultivated area 142 million ha of the net sown area, about 60 per cent requires some soil conservation measures. Certain measures have already been taken in this regard and positive results are now becoming apparent. Better land management practices are required at different levels. This is also necessary because good agricultural lands are encroached upon as a direct result of urbanisation and industrialisation.

There is no national water policy as such, though it is so closely related to the agricultural based economy. The extremities like floods and droughts have become an annual ordeal. The accentuation of floods and damages are mainly attributed to greater environmental degradation, particularly deforestation, unregulated settlements and developments particularly in the flood plains. It is estimated that losses from floods annually amount to Rs. 2,104 million affecting 7.4 million ha of area of which nearly 50 per cent comprises cropland and causing suffering to 50 million persons. On the other extreme, end of hydrological cycle is the drought. At the end of the 6th national development plan, Rs. 337.41 crores were spent on drought prone area programme (DPAP). In the 7th Plan, DPAP projects area being implemented in districts like Udhampur, Almora, Pithoragarh, Pauri, Garhwal and Tehri Garhwal — all in the Himalayas. On the other hand, the cities in the plains and elsewhere are facing water crisis particularly in the summers. The existing municipal or corporation water supply systems are over burdened with expanding population. In the rural scene, there are about 73.6 per cent village electrified but only 20 per cent of the villages with protected water supply.

India is rich in resources. Since Independence, exploration and exploitation of mineral and petroleum resources received due attention but least consideration to the environmental aspects. Energy consumption is very low in comparison to other countries and about 45 per cent is derived from non-commercial sources, such as fuel food, animal and agricultural water. But the demand for energy and mineral resources is increasing due to local consumption and exports. The local production of petrol and petroleum products is increasing continuously. But the use of petrol is always on the increase due to multiplication of different types of vehicles on the road. In expanding cities, housing problem and relatively economic prosperity have contributed to high mobility which generates higher vehicular traffic.

Exploitation of resources, traffic, industrialisation, population beyond carrying capacity and like have lead to pollution of all sorts. In some cities it is becoming alarming, while other cities or their parts are identified with particular type of pollution. The rural areas are also being influenced by pollutants. This is not only due to the industrialisation of the rural area, but also due to the transportability of such elements from urban areas. It is a healthy sign that there is all round concern for environment. Threat to Taj Mahal or disaster in Bhopal or even forest coverage and underground species are matter of national concern.

Retrospection

India is full of contradictions. It has its share of resources and problems. Large country like India with huge resources is in a position to meet any challenge. Its population is a resource if it can be utilised properly. It has one of the largest scientific and technical man power. There have been advancements have been in every field of the national life. It runs one of the largest railway network in the world. Television stations have been opened in all the parts of the country. The satellites are designed and manufactured in the country. It has one of the largest armies equipped with indigenous inter continental ballistic missiles. Its technological achievements have been additional component in contributing to the unity in diversity.

India has the largest number of cattle population in the world but the milk products are not sufficient to meet the demands. It is one of the largest producers of rice, wheat, sugar and oil seeds, but in some years it has to import these materials. Similarly, with all its skilled manpower, items like machinery, iron and steel and chemicals are high in the import list. There are other issues as well where the solution seems to be in the country. It appears that such contradictions can be minimised if a more determined effort is made in planning and development. The issues like environment *vs*. development, population growth and quality of life can be tackled with greater managerial efficiency in the affairs of the nation. India has all the potentialities to regain its past glory. It has already demonstrated its capacity to face the challenges from time to time. The problems coming in the way only determines the path to be chosen in the process of development.

REFERENCES

Adhikari, Sudeepta (1985), India's national integration: A problem in political geography, *Geographical Review of India*, Vol. XLVII (1), pp. 83-90.

Bose, A, *et al.* (eds), (1974), *Population in India's Development*, 1947-2000, Vikas, New Delhi.

Banerjee, B. (1984), Changing cultural scenario and economic development in India, *Geographical Review of India*, Vol. XLVI (2), pp. 55-57.

ICCSR, (1984), *A Survey of Research in Geography, 1972-75*, Concept Publishing Company, New Delhi.

Khoshroo, T.N. (1988), *Environmental Priorities in India and Sustainable Development*, Indian Science Congress Association, New Delhi.

Sethna, H.N. (1977), *Survey, Conservation and Utilisation of Resources*, Presidential Address, Indian Science Congress Association, Calcutta.

APPENDICES

APPENDIX I

INDIA AT A GLANCE

Area : 32,87,263 square kilometre
Population : 68,51,84,692 persons (including projected population of Assam) - 15.69 per cent of the world's population
Extent : 8° 4′ - 37° 6′ N
68° 7′ - 97° 25′ E
(With Nicobar Islands the southern extent is 6° 45′ N)
North-South – 3,214 km
East-West – 2,933 km
Land frontier : 15,200 km
Coastline : 6,100 km for mainland
5,516.6 km with islands
Extent of Himalayas : 2,400 km in length
240-320 km in width
Ganga plain : 2,400 km in length
240-320 in width
First census : 1872 A.D.
Population increase (1971-81) : 25 per cent
Proportion of rural population : 76.69 per cent
Proportion of main workers : 33.45 per cent
Density : 216 persons per square kilometre
Sex-ratio : 933 females per 1,000 males
Literacy : 36.23 per cent
Languages : 15 (as per the 8th schedule of the constitution)
Birth rate : 32.4 per 1,000 population
Death rate : 11.1 per 1,000 population

Expectation of life at birth : 52.5 years for males
52.1 years for females
Mean age at marriage : 23.3 for males
18.3 for females
Annual rate of population increase: 2.5 per cent
Scheduled caste and scheduled tribe population : 23.51 per cent
Towns : 3,949 excluding Assam
Number of million cities : 12
Inhabited villages : 5,57,137 excluding Assam (1981)
21,995 in Assam (1971)
Uninhabited villages : 48,087 excluding Assam (1981)
1,106 in Assam (1971)
States : 25
Union Territories : 7
Districts : About 470

Date of Independence : 15 August, 1947
National Calendar : Based on Saka era with Chaitra as the first month.
National animal : Majestic tiger, Panthera tigris
National Bird : Indian peacock, Pavo cristatus
Total road length : 1,772 thousand km
Villages connected with all weather roads : 2,35,000
Railway routes : 618 thousand kms
Post offices - Urban : 15,600
Rural : 1,28,800
Telegraph offices : 37,600
Educated manpower : 47,724,000
Number of factories : 96,706
Electricity generated : 201.9 Bn. kWh
Villages electrified : 4,26,323
Forest cover : 642 thousand square kilometre
Irrigated area : 113.3 million ha
Availability of foodgrain : 466 grams/capita/day
All types of vehicles on road (1986) : 10,226,000
India's share in world tourism (1986) : 0.43 per cent

India's rank in the world:

Feature	Rank
Area	7
Population	2
Gross domestic product (GDP)	12
Value added in agriculture	4
Value added in manufacturing	16
Electricity generated	11
Arable land	3
Irrigated area	2
Tractors in use	14
Cattle population	1
Rice production	2
Wheat production	4
Groundnut (Shell) production	1
Cotton (lint) production	4
Tobacco production	3
Tea production	1
Potato production	5
Milk (cow & buffalo)	3
Butter and ghee	2
Eggs (hen)	6
Fish catches	7
Sugar	4
Cotton Yard	3
Cement	6
Iron ore	5
Railway passenger km.	3
Railway freight tonne km.	5
Civil aviation	14
Radio receivers	8
Long films production	1
Cinema houses (fixed)	3

Source: Different publications.

APPENDIX II

YEARS OF DROUGHT IN INDIA

Year	Area affected (× 10^6 sq. km)	% area of the country	DI value	Category	Ranking
1876	0.49	15.8	0.98	S	34
1877	2.03	64.7	3.38	C	2
1883	1.03	32.8	1.24	M	13
1884	0.70	22.2	0.60	S	26
1885	0.48	15.4	0.15	S	35
1891	1.15	36.7	1.54	M	9
1896	0.68	21.7	0.57	S	27
1899	1.99	63.4	3.31	C	3
1901	0.89	28.5	1.01	M	20
1902	0.54	17.1	0.27	S	33
1904	0.98	31.1	1.18	M	16
1905	1.09	34.7	1.41	M	10
1907	0.85	27.2	0.93	S	22
1911	0.97	30.8	1.15	M	17
1913	0.70	22.3	0.60	S	25
1915	0.63	20.2	0.47	S	30
1918	2.16	68.7	3.64	C	1
1920	1.22	38.8	1.69	M	8
1925	0.80	25.5	0.81	S	24
1928	0.67	21.4	0.55	S	28
1936	0.86	27.6	0.95	S	21
1941	0.01	32.3	1.26	M	15
1951	1.04	33.2	1.32	M	11
1952	0.81	25.8	0.83	S	23
1965	1.35	42.9	1.95	M	6
1966	1.01	32.3	1.26	M	14
1968	0.65	20.6	0.50	S	29
1969	0.62	19.9	0.45	S	31
1971	0.42	13.3	0.02	S	36
1972	1.39	44.4	2.05	Sv	5
1974	0.92	29.3	1.06	M	19
1979	1.24	39.4	1.72	M	7
1982	1.04	33.1	1.31	M	12
1985	0.95	30.1	1.12	M	18
1986	0.60	19.0	0.39	S	32
1987	1.55	49.2	2.37	Sv	4

DI - Drought Index, DI = $[(x - \bar{x})\ ]$
σ - Standard deviation (0.48 × 10^6 sq. km, for this series).
X - mean (0.41 × 10^6 sq. km or 13.1% of the total area).
S - Slight (0.1 — 1.0), M-Moderate (1.1-2.0), Sv-Severe (2.1 — 3.0)
C - Calamitous (> 3.0)

Source: A. Chowdhury, M.M. Dandekar & P.S. Raut, Variability on drought incidence over India — A statistical approach. *Mausam*, 1989, Vol. 40, No. 2, pp. 207-14.

APPENDIX III

PRODUCTION OF COAL, IRON ORE AND CEMENT

(in million tonnes)

Items	April-March		
	1986-87	1987-88	1988-89 (P)
A.Coal (i + ii)	165.69 (+7.4)	179.85 (+8.5)	194.21 (+8.0)
(i) Coking Coal	38.34 (+9.0)	40.03 (+4.4)	42.05 (+5.0)
(ii) Non-Coking Coal	127.35 (+6.9)	139.82 (+9.8)	152.16 (+8.8)
(b) Lignite	7.21 (-10.3)	8.49 (+17.8)	8.28 (-2.5)
B.Iron Ore	52.3 (+16.2)	50.9 (-2.7)	48.9 (-3.9)
C.Cement	36.59 (+10.4)	39.55 (+8.09)	44.20 (+11.75)

Note: (1) P — Provisional
(2) Figures in brackets represent variations over the previous years.

Source: Report on Currency and Finance, 1988-89, Reserve Bank of India.

APPENDIX IV

DEVELOPMENT OF IRRIGATION POTENTIALS

(in million hectares)

Item	Ultimate	Created up to VI Plan (1984-85)	VII Plan 1985-86 to 1989-90 Target	1985-86 (Achievement)	1986-87 (Achievement)	1987-88 (Antici-pated Achieve-ment)	1988-89 (Target)
1	2	3	4	5	6	7	8
A. Major & Medium	58	30.0	4.3	0.5	0.5	0.7	1.0
B. Minor Irrigation	55	37.5	8.6	1.5	1.6	1.6	2.9
(a) Surface Water	15	9.7	1.5	0.3	0.3	0.3	0.4
(b) Ground Water	40	27.8	7.1	1.2	1.3	1.3	2.5
Total : (A + B)	**113**	**67.5**	**12.9**	**2.0**	**2.1**	**2.3**	**3.9**

***Source*: Ministry of Water Resources, Government of India.**

APPENDIX V

FINANCIAL OUTLAY ON MAJOR, MEDIUM AND MINOR IRRIGATION

(*Rs. Crores*)

Item	During Plan period up to end of VI Plan (1980-81 to (1985-86)	VII Plan (1985-86 to 1989-90) (Target)	1985-86 (Actual)	1986-87 (Actual)	1987-88 (Anticipated)	1988-89 (Target)
1	2	3	4	5	6	7
A. Major & Medium	15060	11556	1880	2169	2231	2583
B. Minor Irrigation	8230	6305	994	1173	1341	1504
of which :						
(a) Government Sector	4492	2805	493	588	638	731
(b) Institutional Sector	3738	3500	501	585	703	773
Total (A + B)	**23290**	**17861**	**2874**	**3342**	**3572**	**4087**

Source: Ministry of Water Resource, Government of India.

APPENDIX VI

AGRICULTURAL PRODUCTION

Commodities	Unit	1970-71	1980-81	1983-84	1984-85	1985-86	1986-87	1987-88	1988-89		1989-90 Target
									Target	Latest estimate	
1	2	3	4	5	6	7	8	9	10	11	12
Rice	Million tonnes	42.2	53.6	60.1	58.3	63.8	60.6	56.4	68.0	70.6	72.5
Wheat	Million tonnes	23.8	36.3	45.5	44.1	47.0	44.3	45.1	52.3	54.6	54.0
Coarse grains	Million tonnes	30.6	29.1	33.9	31.2	26.2	26.8	25.9	33.0	32.2	33.8
Pulses	Million tonnes	11.8	10.6	12.9	11.9	13.4	11.7	11.0	13.3	14.2	14.7
Total foodgrains	**Million tonnes**	**108.4**	**129.6**	**152.4**	**145.5**	**150.4**	**143.4**	**138.4**	**166.6**	**171.6**	**175.0**
Kharif foodgrains	Million tonnes	68.9	77.7	89.2	84.5	85.2	80.2	73.9	92.5	96.5	98.0
Rabi foodgrains	Million tonnes	39.5	51.9	63.2	61.0	65.2	63.2	64.5	74.1	75.1	77.0
Sugarcane (in terms of cane)	Million tonnes	126.4	154.2	174.1	170.3	170.6	186.1	196.7	195.0	204.6	212.0
Cotton	Millions bales of 170 kg each	4.8	7.0	6.4	8.5	8.7	6.9	6.4	9.8	8.7	10.0
Jute & Mesta	Million bales of 180 kg each	6.2	8.2	7.7	7.8	12.6	8.6	6.8	9.2	7.7	9.5
Oilseeds	Million tonnes	9.6	9.4	12.7	12.9	10.8	11.3	12.4	15.7	18.2	16.5
Tea	Million kg	419.0	569.6	581.5	639.9	656.2	624.6	674.3	734.0	700.3	760.0
Coffee	Million kg	109.8	118.6	105.0	195.1	122.3	192.3	123.0	171.0	215.0	180.0

***Source*:** Ministry of Agriculture, Government of India, Tea Board and Coffee Board.

APPENDIX VII

TARGETS AND ACTUAL PRODUCTION OF FOODGRAINS

(*Million tonnes*)

	Rice	Wheat	Coarse-grains	Pulses	Total Foodgrains
1	2	3	4	5	6
1983-84					
Target	57.0	41.0	31.0	13.0	142.0
Actual	60.1	45.5	33.9	12.9	152.4
1984-85					
Target	61.5	45.6	33.5	13.0	153.6
Actual	58.3	44.1	31.2	11.9	145.5
1985-86					
Target	63.5	49.2	33.0	13.5	159.2
Actual	63.8	47.0	26.2	13.4	150.4
1986-87					
Target	65.0	49.0	32.0	14.0	160.0
Actual	60.0	44.3	26.8	11.7	143.4
1987-88					
Target	64.0	50.0	32.0	14.0	160.0
Actual	56.4	45.1	24.9	11.0	138.4
1988-89					
Target	68.0	52.3	33.0	13.3	166.6
Anticipated Achievement	70.6	54.6	32.2	14.2	171.6
1989-90					
Target	72.5	54.0	33.8	14.7	175.0

Source: Ministry of Agriculture, Government of India.

APPENDIX VIII

CRUDE BIRTH AND DEATH RATES, IN INDIA 1901-86

Year	Birth rates	Death rates
1901-11	49.2	42.6
1911-21	48.1	47.2
1921-31	46.4	36.3
1931-41	45.2	31.2
1941-51	39.9	27.4
1951-61	41.7	22.8
1961-71	41.2	19.0
1971-81	37.2	15.0
1984-86	33.1	11.8

Source*: *A Handbook of Population Statistics, Census of India, 1988.

APPENDIX IX

CRUDE BIRTH AND DEATH RATES

States/Union Territories	Crude birth rate 1987	Crude birth rate 1987	Natural increases 1987	Population Doubling time in years
Andhra Pradesh	30.3	9.9	20.4	34
Arunachal Pradesh	36.3	13.2	23.1	30
Assam	34.2	11.6	22.6	30
Bihar	36.6	13.1	23.5	29
Goa	18.9	7.5	11.4	61
Gujarat	30.8	9.8	21.0	33
Haryana	34.5	8.8	25.7	27
Himachal Pradesh	30.8	8.5	22.3	31
Jammu & Kashmir	31.0	7.7	23.3	29
Karnataka	28.9	8.7	20.2	34
Kerala	21.7	6.1	15.6	44
Madhya Pradesh	36.4	13.3	23.1	30
Maharashtra	28.9	8.3	20.6	33
Manipur	25.9	5.6	20.3	34
Meghalaya	34.9	9.1	25.8	27
Nagaland	21.7	4.4	16.8	41
Orissa	31.0	13.1	17.9	38
Punjab	28.7	8.1	20.6	33
Rajasthan	35.1	11.6	23.5	29
Sikkim	33.3	10.3	23.0	30
Tamil Nadu	24.0	9.9	14.1	49
Tripura	28.2	9.2	19.0	36
Uttar Pradesh	37.9	14.5	23.4	29
West Bengal	30.7	8.8	21.9	31
Andaman & Nicobar Is.	26.9	6.1	20.8	33
Chandigarh	23.9	3.8	20.1	39
Dadra & Nagar Haveli	35.8	11.3	24.5	28
Delhi	30.4	7.5	22.9	30
Lakshadweep	30.5	6.3	24.2	28
Pondicherry	22.4	8.0	14.4	48
India	**32**	**10.9**	**21.3**	**32**

***Source*: Sudesh Nangia, India — Population Data sheet, 1989, J.N.U., New Delhi.**

APPENDIX X

MAJOR RELIGIONS, 1981

(*Percentage to total Population*)

State/Union Territories	Hindu	Muslim	Christians	Sikhs	Buddhists	Jain	Other religion	N.K.
1	2	3	4	5	6	7	8	9
Andhra Pradesh	88.75	8.47	2.68	0.03	0.02	0.03	-	0.02
Arunachal Pradesh	29.24	0.80	4.32	0.19	13.69	0.01	51.60	0.51
Bihar	82.97	14.13	1.06	0.11	-	0.04	1.09	-
Goa	65.90	4.46	29.28	0.13	-	-	-	-
Gujarat	89.53	8.53	0.39	0.07	0.02	1.37	0.05	0.04
Haryana	89.36	4.05	0.09	6.21	0.01	0.27	0.01	-
Himachal Pradesh	95.77	1.63	0.09	1.22	1.23	0.02	0.01	0.03
Jammu & Kashmir	32.24	64.19	0.14	2.23	1.17	0.03	-	-
Karnataka	85.77	11.21	2.08	0.02	0.11	0.77	0.04	-
Kerala	58.15	21.25	20.56	0.01	-	0.02	-	0.01
Madhya Pradesh	92.96	4.80	0.68	0.27	0.14	0.85	0.30	-
Maharashtra	81.40	9.25	1.27	0.17	6.28	1.50	0.12	0.01
Manipur	60.04	6.99	29.68	0.07	0.03	0.07	2.50	0.02

Contd.

Contd.

1	2	3	4	5	6	7	8	9
Meghalaya	18.03	3.10	52.62	0.13	0.02	0.04	25.77	0.11
Mizoram	7.14	0.45	83.81	0.09	8.19	-	0.32	-
Nagaland	14.36	1.52	80.21	0.10	0.07	0.15	3.59	-
Orissa	95.42	1.60	1.82	0.05	0.03	0.03	1.04	0.01
Punjab	36.93	1.00	1.10	60.75	-	0.16	0.05	0.01
Rajasthan	89.32	7.28	0.12	1.44	0.01	1.82	0.01	-
Sikkim	69.25	1.03	2.22	0.10	28.71	0.03	0.63	0.03
Tamil Nadu	88.86	5.21	5.78	0.01	-	0.10	0.40	-
Tripura	89.34	6.75	1.21	0.01	26.7	0.02	-	-
Uttar Pradesh	83.31	15.93	0.55	0.41	0.05	0.13	0.02	-
West Bengal	76.96	21.51	0.59	0.09	0.29	0.07	0.48	0.01
Andaman & Nicobar Is.	64.53	8.58	25.58	0.52	0.07	-	0.12	0.60
Chandigarh	75.27	2.02	0.99	21.11	0.10	0.42	0.06	0.03
Dadra & Nagar Haveli	95.96	1.86	1.95	0.01	0.18	0.36	0.07	0.01
Delhi	83.60	7.75	0.99	6.33	0.11	1.19	0.02	0.01
Daman & Diu	70.90	4.46	24.28	0.13	0.03	0.06	0.05	0.09
Lakshadweep	4.47	94.84	0.66	-	-	-	-	0.03
Pondicherry	84.97	6.06	8.26	-	0.01	0.05	0.03	0.02
India	**82.63**	**11.36**	**2.43**	**1.96**	**0.71**	**0.48**	**0.42**	**0.01**

N.K. — Not Known

***Source*: Census of India, 1981**

APPENDIX XI

PROFILES OF FOUR METROPOLITAN CITIES

	Bombay (Greater)	Calcutta (U.A.)	Delhi (U.A.)	Madras (U.A.)
Area (Sq. km.)	603	852	541	572
Population ('000s)	8,243	9,194	5,729	4,289
Males ('000s)	4,653	5,162	3,169	2,222
Females ('000s)	3,591	4,032	2,561	2,067
Females per 1,000 males	772	781	808	930
Population per sq km ('000s)	13.7	10.8	10.6	7.5
Literate population ('000s)	5,620	6,020	3,595	2,892
Males ('000s)	3,439	3,661	2,190	1,670
Females ('000s)	2,181	2,359	1,404	1,222
Literacy rate (%)	68.2	65.5	62.7	67.4
Males (%)	73.9	70.9	69.1	75.2
Females (%)	60.7	58.5	54.8	59.1
Working population ('000s)	2,860	2,800	1,846	1,208
Males ('000s)	2,553	2,616	1,678	1,068
Females ('000s)	308	183	167	140
Working population ratio (%)	34.7	30.5	32.2	28.2
Males (%)	54.9	50.7	53.0	48.1
Females (%)	8.6	4.5	6.5	6.8
Occupied houses ('000s)	1,591	1,721	1,081	822
Households ('000s)	1,625	1,747	1,133	834
Houseless population ('000s)	44	64	23	8
Slum dwellers ('000s)	2,831	3,028	1,800	1,363
Cargo handled ports (Mn.t)	25.1	4.1	*Nil*	19.8

Contd.

Contd.

	Bombay (Greater)	Calcutta (U.A.)	Delhi (U.A.)	Madras (U.A.)
International airport traffic				
Passengers handled ('000s)	8,562	2,229	5,572	1,757
Cargo handled ('000s tonnes)	169	33	107	35
Motor vehicles ('000s)	441	286	841	139
Road accidents per 10,000 motor vehicles:				
Accidents (No.)	607	279	74	417
Persons killed (No.)	15	15	15	28
Persons injured (No.)	181	106	76	310
Per capita drinking water availability per day (liters)	112/135	259	239	73
Telephone exchanges ('000s)* (incl. Private branch exchanges)	6.3	3.5	2.6	1.7
Telephones ('000s)*	652	299	426	160
Hospital beds per 1,000 population (No.)	3.3	4.1	2.1	3.2

Note: Figures up to 'Slum dwellers' relate to 1981; motor vehicles, road accidents and hospital beds to 1984-85 or 1983-84; telephones and exchange to 1985-86 and the rest to 1986-87 U.A. = Urban Agglomeration.
* Relates to district covered by each city.

Source: Statistical Outline of India.

APPENDIX XII

PERFORMANCE OF DISTRICT INDUSTRIES CENTRES

Item	Total Achievements (April-March)			Per DIC Achievements (April-March)		
	1986-87 (Actual)	1987-88 (Estimated)	1988-89 (Estimated)	1987-87 (Actual)	1987-88 (Estimated)	1988-89 (Estimated)
1	2	3	4	5	6	7
1. No. of DICs sanctioned	419	422	422	—	—	—
2. No. of DICs reporting	418	422	422	—	—	—
3. New Registrations done (Nos.)	374,023	385,286	392,953	895	913	931
4. New Units established (Nos.)						
(i) Artisans	316,169	328,316	337,673	756	778	800
(ii) SSI	106,211	112,252	117,719	254	266	279
Total	422,380	440,568	455,392	1,010	1,044	1,079
5. Credit assistance Provided by financial institutions through DIC (Rupees in crores)	890.80	1371.50	2094.69	2.13	3.25	4.96
6. Additional employment generated (No. of persons)	1333,024	1411,168	1479,892	3,189	3,344	3,507
7. Units given technical and other assistance (Nos.)	30,810	64,500	96,800	74	153	229

Source: Office of the Development Commissioner, Ministry of Industry, Government of India, New Delhi.

APPENDIX XIII

PRODUCTION OF STEEL

(Thousand tonnes)

Year (April-March)	Bhilai	Durgapur	Rourkela	Bokaro	TISCO	IISCO	Total (2 to 7)	VISL	ASP	Total (9+10)
1	2	3	4	5	6	7	8	9	10	11
Production of Hot Metal										
1986-87	2,510	1,125	1,223	2,813	1,941	825	10,437 (+4.0)	5	—	5 (-84.8)
1987-88	2,556	1,138	1,212	3,123	2,018	818	10,865 (+4.1)	—	—	—
1988-89 P	3,306	1,096	1,252	3,221	2,239	768	11,882 (+9.4)	12	—	12
Production of Saleable Steel										
1986-87	2,150	751	1,140	1,745	1,907	526	8,219 (+5.7)	52	82	134 (+16.5)
1987-88	2,173	835	1,156	1,968	1,914	542	8,568 (+4.5)	40	89	129 (-3.7)
1988-89 P	2,542	832	1,168	2,277	1,945	442	9,206 (+7.2)	41	107	148 (+14.7)
Production of Crude Steel										
1986-87	2,230	922	1,100	2,056	2,250	528	9,086 (+6.1)	56	140	196 (+14.6)
1987-88	2,471	936	1,115	2,418	2,277	546	9,763 (+7.5)	53	154	207 (+5.6)
1988-89 p	3,094	956	1,190	2,771	2,314	465	10,790 (+10.5)	58	179	237 (+14.5)

P : Provisional. VISL: Visvesvaraya Iron and Steel Ltd. ASP: Alloy Steel Plant.
Note: Figures in brackets represent percentage variations over the previous year.
Source: Steel Authority of India Ltd.

APPENDIX XIV

CRUDE OIL AND PETROLEUM REFINERY PRODUCTS

(Thousand Tonnes)

Item	April-March		
	1986-87	1987-88	1988-89 (P)
1	2	3	4
A. *Crude Oil*			
(a) Production (i+ii)	**30,480**	**30,357**	**32,040**
	(+1.0)	(-0.4)	(+5.5)
(i) On-shore	9,862	10,193	10,928
	(+5.5)	(+3.4)	(+7.2)
(ii) Off-shore	20,618	20,164	21,112
	(-1.0)	(-2.2)	(+4.7)
(b) Imports	**15,476**	**17,732**	**17,811**
	(+2.2)	(+14.6)	(+0.4)
B. *Petroleum Refinery Products*	42,761	44,728	45,597
(a) Production of which	(+6.3)	(+4.6)	(+1.9)
(i) Naphtha	5,437	5,462	5,372
	(+9.7)	(+0.5)	(-1.6)
(ii) Kerosene	4,912	5,104	5,202
	(+21.9)	(+3.9)	(+1.9)
(iii) High Speed Diesel Oil	15,450	16,296	16,656
	(+5.6)	(+5.5)	(+2.2)
(b) Imports of which	**3,047**	**3,948**	**6,258**
	(-21.2)	**(+29.6)**	**(+58.5)**
(i) Naphtha	34	—	—
	—	—	—
(ii) Kerosene	1,900	2,181	2,747
	(-26.0)	(+14.8)	(+26.0)
(iii) High Speed Diesel Oil	750	1,329	2,448
	(-15.7)	(+77.2)	(+84.2)

P: Provisional.

Note: Figures in brackets represent percentage variations over the preceding year.

Source: Ministry of Petroleum and Natural Gas.

APPENDIX XV

PRODUCTION IN DIFFERENT INDUSTRIES

Items	Unit	April-March 1986-87	1987-88	1988-89
1		2	3	4
A. *Heavy Mechanical Eng.*				
1. Automobiles				
(i) Passenger car	'000 Nos.	100.4 (+14.6)	138.3 (+37.7)	154.7 (+11.9)
(ii) Commercial Vehicles	'000 "	106.0 (+2.9)	119.9 (+13.1)	118.8 (-0.9)
(iii) Jeeps	'000 "	29.3 (+2.1)	23.6 (+19.5)	36.3 (+8.0)
2. Two-wheelers	'000 "	1,434.7 (+16.7)	1,596.9 (+11.3)	1,674.5 (+4.4)
3. Structurals	'000 "	129.0 (-2.3)	131.0 (+1.6)	107.8 (-17.7)
4. Steel Castings	'000 "	238.2 (+4.8)	311.4 (30.7)	332.7 (+6.8)
5. Diesel Engines	'000 "	2,235.7 (+8.9)	1,896.6 (-15.2)	1,856.2 (-2.1)
6. Railway Wagons (in terms of 4 wheelers)	'000 "	15.2 (+16.0)	15.8 (+3.9)	21.7 (+37.3)
7. Road Rollers	'000 "	662 (-17.1)	530 (-14.8)	544 (+3.6)
8. Lifts	'000 "	1,880 (+24.5)	1,952 (+3.8)	2,210 (+13.2)
9. Tractors	'000 "	80.0 (+4.8)	82.9 (+3.6)	109.7 (32.3)
B. *Light Mechanical Eng.*				
1. Type writers	'000 "	122.9 (+1.2)	93.6 (-23.8)	122.0 (30.3)
2. Bicycles	'000 "	6,118.6 (+10.2)	6,706.5 (+9.6)	6,702.7 (-0.7)
3. Sewing Machines	'000 "	377.1 (+29.4)	314.6 (-16.6)	238.3 (-24.3)

(Contd.)

(Contd.)

1		2	3	4
4. Razor Blades	Mill Nos.	2,108.4 (+3.4)	2,235 (+6.0)	2,711.5 (+21.3)
5. Wrist watches	Mill Nos.	7.54 (+41.5)	7.50 (-0.5)	9.00 (+20.0)
C. *Light Electrical Engineering*				
1. Electrical Lamps				
(i) Incandescent	Mill Nos.	283.7 (+4.9)	255.5 (-9.9)	247.5 (-3.1)
(ii) Flourescent	Mill Nos.	47.0 (+10.3)	48.7 (+3.6)	46.2 (-5.1)
2. Electrical Fans	Mill Nos.	5.3 (-17.2)	6.1 (+15.1)	6.4 (+4.4)
3. Radio Receivers	'000 "	1,248.8 (+7.5)	977.9 (-21.7)	1,035.9 (+5.9)
4. Storage Batteries	'000 "	2,601.0 (+17.1)	2,933.0 (+12.8)	3,284.7 (+12.0)
5. Dry cells	Mill Nos.	1,167.0 (-1.8)	1,700.2 (+45.7)	1,352.0 (-20.5)
6. Domestic Refrigerators	'000 "	590.5 (-12.9)	680.3 (-15.2)	945.7 (+39.0)
D. *Heavey Electrical Engineering*				
1. Power Transformers	Mill KVA	28.2 (+3.3)	24.21 (-14.5)	28.6 (+18.7)
2. Electric Motors	Mill H.P.	5.4 (+3.8)	4.3 (-20.4)	5.4 (+25.6)
3. Power Driven pumps	'000 Nos.	459.4 (-10.3)	515.7 (+12.3)	593.8 (+15.1)

Source: Report on Currency and Finance, 1988-89, Reserve Bank of India.

APPENDIX XVI

SOURCE-WISE BREAK-UP OF ELECTRIC POWER

(Million Kilo Watt Hours)

Source	April-March					
	1986-87		1987-88		1988-89 (P)	
	Power Genera-tion	Percen-tage Share	Power Genera-tion	Percen-tage Share	Power Genera-ration	Percen-tage Share
1	2	3	4	5	6	7
1. Thermal	128,818 (+12.9)	68.7	149,464 (+16.0)	74.0	157,510 (+5.4)	71.2
2. Hydel	53,764 (+5.6)	28.6	47,396 (-11.8)	23.5	57,793 (+21.9)	26.2
3. Nuclear	5,023 (+0.8)	2.7	5,034 (+0.2)	2.5	5,822 (+15.7)	2.6
4. Total	187,605 (+10.3)	100.00	201,894 (+7.6)	100.00	221,125 (+9.5)	100.00

P: Provisional.
Note: Figures in brackets represent variations over the previous year.
Source: Central Electricity Authority.

APPENDIX XVII

PROVISIONAL POPULATION TOTALS

A COMPARIATIVE PICTURE OF THE POPULATION

Density : 267 Persons per sq. km. *Sex-ratio 929F/1000M*

India/State/ Union Territory	Total population 1981	Total population 1991	Growth rate 1971-81	Growth rate 1981-91	Density population 1991
2	3	4	5	6	7
INDIA	683,329,097*	843,930,861	+24.66"	23.50	^^267
States					
Andhra Pradesh	53,549,673	66,304,854	+23.10	+23.82	241
Arunachal Pradesh	631,839	858,329	+35.15	+35.86	10
Assam	18,041,248@	22,294,562	+23.36	+23.58	284
Bihar	69,914,734	86,338,853	+24.06	+23.49	497
Goa	1,0?7,749	1,168,622	+26.74	+15.96	316
Gujarat	34,085,799	41,174,060	+27.67	+20.80	210
Haryana	12,922,119	16,317,715	+28.75	+26.28	369
Himachal Pradesh	4,280,818	5,111,079	+23.71	+19.39	92
Jammu & Kashmir	5,987,389	7,718,700#	+29.69	+28.92	^^76
Karnataka	37,135,714	44,817,398	+26.75	+20.69	234
Kerala	25,453,680	2,90,11,237	+19.24	+13.98	747
Madhya Pradesh	52,178,844	66,135,862	+25.27	+26.75	149
Maharashtra	62,784,171	78,706,719	+24.54	+25.36	256
Manipur	1,420,953	1,826,714	+32.46	+28.56	82
Meghalaya	1,335,819	1,760,626	+32.04	+31.80	78
Mizoram	493,757	686,217	+48.55	38.98	33
Nagaland	774,930	1,215,573	+50.05	+56.86	73
Orissa	26,370,271	31,512,070	+20.17	+19.50	202
Punjab	16,788,915	20,190,795	+23.89	+20.26	401

Contd.

Contd.

2	3	4	5	6	7
Rajasthan	34,261,862	43,880,640	+32.97	+28.07	128
Sikkim	316,385	403,612	+50.77	+27.57	57
Tamil Nadu	48,408,077	55,638,318	+17.50	+14.94	428
Tripura	2,053,058	2,744,827	+31.92	+33.69	262
Uttar Pradesh	110,862,512	138,760,417	+25.49	+25.16	471
West Bengal	54,580,647	67,982,732	+23.17	+24.55	766
Union Territories					
A & N Islands	188,741	277,989	+63.93	+47.29	34
Chandigarh	451,610	640,725	+75.55	+41.88	5,620
Dadra & Nagar Haveli	103,676	138,542	+39.78	+33.63	282
Daman & Diu	78,981	101,439	+26.07	+28.43	906
Delhi	6,220,406	9,370,475	+53.00	+50.64	6,319
Lakshadweep	40,249	51,681	+26.53	+28.40	1,615
Pondicherry	604,471	789,416	+28.15	+30.60	1,605

@ In 1981, census was not conducted in Assam. Based on the 1971 Census population and the 1991 Census provisional resutls, the population of Assam for 1981 has been interpolated.

* As a consequence of the revised for Assam for the year 1981, the total population of India as of 1981 has been estimated as 683,329,097 as against earlier published figuresof 685,184,692.

The 1991 Census has not yet been conduced in Jammu and Kashmir. The figures are as projections prepared by the Standing Committee of Experts on Population Projections, October, 1989.

" As a consequence of the revised figures, the decadal growth rate for India during 1971-81 has been estimated as 24.66 per cent.

^^ Density has been worked out on comparable data.

BIBLIOGRAPHY

Agarwala, S.N., *India's Population Problem*, McGraw-Hill Publication, Delhi, 1977.

Akhtar, Rais & A.T.A. Learmonth, *Geographical Aspects of Health & Diseases*, Concept, New Delhi, 1986.

Bose, A (ed.), *Patterns of Population Change in India, 1951-61*, Allied Publishers, Bombay, 1967.

— *India's Urbanisation, 1901-2001*, Tata McGraw-Hill, New Delhi, 1978.

— *India's Urbanisation, 1901-2001*, Second edition, Tata McGraw-Hill, New Delhi, 1980.

— *From Population to People*, Vol. I & II B.R. Publishing Corporation, Delhi.

Chandrasekhar, C.S. & Deva Raj (eds.) *Urban Perspective 2001*, National Institute of Urban Affairs, New Delhi, 1918.

Chatterjee, S.P., Himalayan Mountain Ranges, *Encyclopaedia Britannica*, 1974.

Chaudhuri, B. & A.K. Maiti (eds.), *Forest Development in India*, Inter-India, New Delhi, 1989.

Chaudhuri M.R., *Power Resources of India*, Oxford & IBH Publishing Co., Calcutta, 1970.

ESCAP, *Study and Review of the Human Settlements situation in Asia and the Country Monograph*, United Nations, Vol. 2 (Chapter V : India), (Undated).

Karan, P.P. and W.M. Jenkins, *The Himalayan Kingdoms: Bhutan, Sikkim and Nepal*, D. Van Nostrand Co, Inc., Princton, 1963.

Kayastha, S.L., *The Himalayan Beas Basin: A Study in Habitat, Economy and Society*, Banaras Hindu University Press, Varanasi, 1964.

Krishnan, M.S., *Geology of India and Burma*, Madras, 1960.

Kumar, V.K., *Kanpur City: A Study in Environmental Pollution*, Tara Book Agency, Varanasi, 1982.

Mitra, Asok, *et al.*, *Population and Area of Cities, Towns and Urban Agglomerations, 1872-1971*, Allied Publisher, Bombay, 1980.

Mitra, Asok, Shekhar Mukherjee, R. Bose, *Indian Cities: Their Industrial Structure, Immigration and Capital Investment*, Abhinav Publishers, New Delhi, 1980.

Mukherji, A.B. & Aijazuddin Ahmad (eds.), *India: Culture, Society and Economy*, Inter-India, Publication, New Delhi, 1985.

Markanday, Kalpana, *Regional Urban Centres : Structures and Interaction*, Inter-India, New Delhi, 1986.

Mishra, R.P. *et al.*, *Regional Development Planning in India*, New Delhi, 1976.

Nag, P., A proposed base for geographical information system for India, *International Journal on Geographical Information System*, Vol. 1, No. 2, 1987.

Nag, P., India, in P. Nag (ed.), *Census Mapping Survey*, Concept Publishing Co., IGU Commission on Population Geography, New Delhi, 1984.

Nag, P., Current trends of historical approaches in Geographical researches in India. *Historika Geographic* — **Historical Geography**, Vol. 19, Institute of Czechoslovak and World History, Czechoslovak Academy of Sciences, Prague, 1980, pp. 25-35.

Nag, P., Global population trend and their geopolitical implications, *Journal of the National Association of Geographers of India*, Vol. IX(2), 1989, pp. 39-48.

NATMO, *Tourist Atlas of India*, National Atlas & Thematic Mapping Organisation, Calcutta, 1974.

—, *Atlas of Forest Resources of India*, National Atlas & Thematic Mapping Organisation, Calcutta, 1976.

—, *Atlas of Agricultural Resources of India*, National Atlas & Thematic Mapping Organisation, Calcutta, 1978.

—, *National Atlas of India*, National Atlas & Thematic Mapping Organisation, Calcutta, (8 Vols.) 1982.

—, *Glimpses of India*, National Atlas & thematic Mapping Organisation, Occasional Monograph 2, Calcutta, 1985.

—, *Irrigation Atlas of India*, National Atlas & Thematic Mapping Organisation, Calcutta (2nd Ed.), 1990.

Pandey, D.C. & P.C. Tiwari (eds.) *Dimensions of Development Planning*, Criterion Publications New Delhi, 1989.

Panduranga Rao, D,. *Trends in Indian Transportation System: A Districtwise Study*, Inter-India, New Delhi, 1985.

Panduranga Rao, D, (ed.), *Problem of Urban Transportation in India*, Inter-India, New Delhi, 1986.

Panikkar, K.M., *Geographical Factors in Indian History*, Bharatiya Vidya Bhavan, Bombay, 1955.

Prakash Rao, V.L.S., *Urbanisation in India: Spatial Dimensions*, Concept, New Delhi, 1983.

Premi, M.K., Pattern of internal migration in India : Some new dimensions, in S.N. Singh, M.K. Premi, P.S. Bhatia & Ashish Bose (ed.), *Population Transition in India*, B.R. Publishing Corporation, Delhi, pp. 277-92, 1989.

Premi, M.K., The demographic situation in India, Paper of the East-West Population Institute, Hawaii, No. 80, 1982.

Prasad, Rajendra, *Politico-Geographical Analysis of the Arthasastra*, Inter-India, New Delhi, 1989.

Puri, G.S., *Indian Forest Ecology*, Vol. I, and II, Oxford Books & Stationary, Delhi, 1950.

Rammohan Roy, R. & S. Simhadri, *Development Dynamics in Command Areas of Major Irrigation* Projects, Inter-India, New Delhi, 1989.

Rao, K.L., *India's Water Wealth*, Orient Longman, Calcutta, 1975.

Ray Chaudhuri, S.P. *et al.*, *Soils of India*, Indian Council for Agricultural Research, New Delhi, 1963.

Ray Chaudhuri, S.P., *Land and Soil*, National Book trust, New Delhi, 1966.

Reddy, N.B.K. (ed), *Drought Prone Areas of India*, Rayalseema Geographical Society, S.V. University, Tirupati, 1979.

Raza, Moonis & Sudesh Nangia, *Atlas of the Child in India*, Concept, New Delhi, 1986.

RGI, *Urban Growth in India, 1951-81 (A Statistical Analysis)*, Census of India Pub., 1981.

—, *General Population Tables*, Series-I India, Part II- A(i), Census of India Pub., 1981.

—, *Standard Urban Area*, Series-I, India Part II-A (III), Census of India, Pub., 1981.

—, *General Population Tables*, Series-I, India, Part II-A (II), Census of India Pub., 1981.

Saxena, P.C., Teaching and research in demography, in S.N. Singh, M.K. Premi, P.S. Bhatia & Ashish Bose (ed.), *Population Transition in India*, B.R. Publication Corporation, New Delhi, pp. 25-35, 1989.

Schwartyberg, J.E., *A Historical Atlas of South Asia*, University of Chicago Press, Chicago, 1978.

Sengupta, Smita, *Residential Pattern of Suburbs*. Concept Publishing Company, New Delhi, 1988.

Sengupta, P., *The India Jute Belt*, Orient Longmans, Bombay, 1959.

Sengupta, P. and G. Sdasyuk, *Economic Regionalisation of India: Problems and Approaches*, Census of India 1961, Monograph No. 8, New Delhi, 1968.

Shafi, Mohammad, *Agricultural Productivity and Regional Imbalances: A Study of Uttar Pradesh*, Concept, New Delhi, 1984.

Sharma, S.C. (ed.), *Utilisation of Wastelands for Sustainable Development in India*, Concept Publishing Company, New Delhi, 1990.

Singh, A.K., V.K. Kumra & J. Singh, *Forest Resources Economy and Environment*, Concept New Delhi, 1987.

Singh, H.P., *Resource Appraisal and Planning in India*, New Delhi, 1979.

Singh, K.N. and D.N. Singh (ed.), *Rural Development in India: Problems, Strategies and Approaches*. NGSI Publication, Varanasi, 1985.

Singh, O.P. (ed.), *The Himalaya : Nature, Man and Culture*, Rajesh Publication, New Delhi, 1983.

Singh, Rana P.B., *Clan Settlements in Saran Plain*, NGSI Publication, Varanasi, 1976.

Singh, R.B., *Geography of Rural Development : The Indian Micro-level Experience*, Inter-India New Delhi, 1986.

Singh, T.V. & J. Kaur (eds.), *Himalayas: Mountain and Men*, Print House (India), Lucknow, 1983.

Singh, R.L. *et al.* (eds.), *India: A Regional Geography*, NGSI Publication, Varanasi, 1971.

Singh, Upendra Kumar, *Ganga Pollution and Health Hazard*, Inter-India, New Delhi, 1986.

Sopher, David E., *An Exploration of India : Geographical Perspectives on Society*, Cornell University Press, Ithaca, 1980.

Spate, O.H. K. & A.T.A Learmonth, *India and Pakistan*, Methun, London, 1967.

Taneja, K.L., *Morphology of India Cities*, NGSI Publication, Varanasi, 1971.

Wadia, D.N., *Geology of India*. Macmillan & Co., London, 1960.

Yadav, Atar Singh, *Changing Political Boundaries in India*, Inter-India, New Delhi, 1987.

INDEX